LOVING + HATING
MATHEMATICS

LOVING + HATING
MATHEMATICS

Challenging the Myths
of Mathematical Life

Reuben Hersh and Vera John-Steiner

PRINCETON UNIVERSITY PRESS   PRINCETON AND OXFORD

Published by Princeton University Press, 41 William Street,
Princeton, New Jersey 08540
In the United Kingdom: Princeton University Press, 6 Oxford
Street, Woodstock, Oxfordshire OX20 1TW
press.princeton.edu

Library of Congress Cataloging-in-Publication Data

Hersh, Reuben, 1927–
Loving and hating mathematics : challenging the myths of mathematical life
/ Reuben Hersh and Vera John-Steiner.
p. cm.
Includes bibliographical references and index.
ISBN 978-0-691-14247-0 (hardcover : alk. paper) 1. Mathematicians—
Biography. 2. Mathematicians—Psychology. 3. Mathematics—Study and
teaching—United States. I. John-Steiner, Vera, 1930– II. Title.
QA28.H48 2011
510.92—dc22   2010014452
[B]

British Library Cataloging-in-Publication Data is available

This book has been composed in Sabon

Printed on acid-free paper. ∞

Printed in the United States of America

1 3 5 7 9 10 8 6 4 2

This book is dedicated to our teachers in New York, Cambridge, Budapest, Geneva, and Chicago, and to a teacher we never met, the courageous Bella Abramovna Subbotovskaya.

CONTENTS

ACKNOWLEDGMENTS

This book would not have been possible without the consistent, intelligent, and enthusiastic support of the librarians at the Santa Fe Institute, Ms. Margaret Alexander and Mr. Timothy Taylor, and the commitment and clarity of thought and language of Valerie Clement. This book greatly benefited from the support and editorial wisdom of Vickie Kearn of Princeton University Press and her staff.

We would like to acknowledge and extend thanks for the contribution of our friend Frank Wimberly, who translated from Spanish to English the words of José Luis Massera and his compatriots about their experiences in the Uruguayan prison called Libertad.

For reading one or several chapters in draft form, for permission to quote, and for correcting errors and verifying accuracy, we thank Roger Frye, Claudia Henrion, Alexander Shen, Freeman Dyson, Ivor Grattan-Guinness, Chandler Davis, Allyn Jackson, Roy Lisker, Jenny Harrison, Moe Hirsch, Calvin C. Moore, Robert Osserman, Underwood Dudley, Nel Noddings, Kristin Umland, Cathy Fosnot, Peter Lax, Mary Ellen Rudin, Clarence Stephens, Nel Kroeger, Laura Cameron, Harry Lucas Jr., Richard J. Griego, Don Lemons, Jim Dudley, Peter Ross, Marjorie Senechal, Richard Kitchen, *The Mathematical Intelligencer*, the American Mathematical Society, and the Mathematical Association of America.

For supportive, instructive, and inspiring conversations and letters over the years, we thank Sergio Albeverio, Michael Baron, Jonathan Borwein, Billy Brown, Mario Bunge, Laura Cameron, Menon Charbonneau, Paul Cohen, Brian Conrey, John Conway, Vageli Coutsias, Chandler Davis, Phil and Hadassah

Davis, Martin Davis, Persi Diaconis, Jim and Mary Dudley, Underwood Dudley, Harold Edwards, Jim Ellison, Pedro Embid, Bernie Epstein, Dick Epstein, Paul Fawcett, Paul Fife, Cathy Fosnot, Marilyn Frankenstein, Claire and Roger Frye, Murray Gell-Mann, Jekuthiel Ginsberg, Sylvia P. Glick, Brian Greer, Richard J. Griego, Tom and Rosa Hagstrom, Liong-Shin Hahn, Cliff Harris, Eva Hersh, Einar Hille, Moe Hirsch, Fritz John, Maria del Carmen Jorge Jorge, Tom, Gael, and Nick Keyes, Allyn Jackson, Kirk Jensen, Richard Kitchen, Morris Kline, Steve Krantz, Serge Lang, Anneli and Peter D. Lax, Don Lemons, Ina Lindemann, Roy Lisker, Jens Lorentz, Wilhelm Magnus, Elena Marchisotto, Lisa Mersky, Cathleen Morawetz, Ed Nelson, Nel Noddings, Paul Noren, Bob Osserman, Cristina Pereyra, Ralph Philips, Klaus Peters, Arthur Powell, Gian-Carlo Rota, Peter Ross, Jill and Neal Singer, J. J. Schaffer, Santiago Simanca, Ernesto Sobrevilla-Soto, Stan Steinberg, Constantino Tsallis, Robert Thomas, Kristin Umland, Greg Unnever, Wilfredo Urbina, Cotten and Larry Wallen, Hao Wang, and Frank Wimberly.

Remaining errors and inaccuracies are of course the responsibility of the authors.

LOVING + HATING
MATHEMATICS

Introduction

This book, unlike most books on mathematics, is about mathematicians, their extraordinary passion for mathematics and their full complexity of being. We emphasize the social and emotional sides of mathematical life.

In the great and famous works of Euclid and Newton, we find axioms and theorems. The mathematics seems to speak for itself. No first person speaks, no second person is addressed: "Here is Truth, here is Proof, no more need be said." Going back to the reports of Plato and Descartes, mathematical thinking has been seen as pure reason—a perfect and eternal faculty. The thoughts, feelings, and tribulations of the mathematician are not included.

But it doesn't take deep reflection to realize that this perfection is a human creation. Mathematics is an artifact created by thinking creatures of flesh and blood. Indeed, this fact has always been known to poets, novelists, and dramatists. Mathematicians, like all people, think socially and emotionally in the categories of their time and place and culture. In any great endeavor, such as the structuring of mathematical knowledge, we bring all of our humanity to the work. Along with reasoning, the work includes the joy of discovery, the struggle with uncertainty, and many other emotions. And it is shaped by social realities, including war, political oppression, sexism, and racism, as they have affected society and mathematicians in different eras.

Today the connection between thought and emotion is a major active field of scientific research. Recently the neuroscientist Antonio Damasio and a collaborator wrote, "Modern biology reveals humans to be fundamentally emotional and social creatures. And yet those of us in the field of education often fail to consider that the high-level cognitive skills taught in schools, including reasoning, decision making, and processes related to language, reading, and mathematics, do not function as rational, disembodied systems, somehow influenced by but detached from emotion and the body . . . hidden emotional processes underlie our apparently rational real-world decision making and learning."[1] While purely rational decision making is possible in highly structured situations, they point out that emotionally empowered thinking is needed for recruiting "these skills and knowledge usefully outside the structured context of school and laboratory."[2]

Indeed, mathematical experience advances between the twin poles of exaltation and despair. Granted, despair is more familiar to novices, and exaltation is more associated with the great discoverers. But these opposite emotions lie in waiting, during every mathematical struggle, at every level. The deep emotional connections between mathematical experience in earliest childhood, in maturity, and in ripe old age are an important theme in this book.

To unveil these different aspects of mathematical life, we have read many biographies and autobiographies. We quote a great number of cherished life stories of mathematicians who both delighted in and laughed at the quirks of living with numbers and abstraction. We often let famous mathematicians speak for themselves.

One of our major goals is to overcome some distorted, stereotypic images of the field and its practitioners.[3]

First is the myth that mathematicians are different from other people, lacking emotional complexity.

There is a common belief that in order to engage in complex abstract reasoning, a researcher must exclude emotion from his or her thinking. Our first four chapters refute this belief. A mathematician, like anyone else, has an emotional life, which is sustained by nurturance in childhood and youth and by companionship and mutual support in later years.

It's a challenge for everyone to achieve balance in one's emotional life. It's a particularly severe challenge for those working in mathematics, where the pursuit of certainty, without a clearly identified path, can sometimes lead to despair. The mathematicians' absorption in their special, separate world of thought is central to their accomplishments and their joy in doing mathematics. Yet all creative work requires support. Enmeshed in a mental world far removed from the understanding of those with whom he or she is closest, a mathematician risks becoming psychologically isolated. We tell of some mathematicians for whom such isolation became extreme and overwhelmed their existence. We also tell of others for whom the clarity and beauty of mathematics provided an emotional haven against persecution or the tragic effects of war. The common thread is loving and hating mathematics.

We start in chapter 1 by asking, How does a child first begin to become a mathematician? We explore the thrill of discovery and the power of engagement that some young people can experience. We listen to young contestants in international mathematical olympiads and to their parents. We follow young mathematicians through their graduate school years. Among the emotions that accompany mathematical activity, we find affinity and doubt, frustration and elation, camaraderie and rivalry, and friendship and jealousy.

The culture of mathematics is the subject of chapter 2. Socialization into a community implies the sharing of values, cognitive approaches, and social beliefs and practices. It also involves ways of handling internal tensions that can rupture much needed human and professional ties. We report on three recent episodes in the history of mathematics that made headlines: the proof of Fermat's Last Theorem; the recognition of the group of phenomena in dynamics known as "chaos"; and the proof of Thurston's program in four-dimensional topology, including the Poincaré conjecture. This involved two great geometers, Grisha Perelman and S.-T. Yau, and shows us the cost of single-minded commitment and the mathematician's ambivalence toward fame. We also report the struggles of the University of California professor Jenny Harrison. (A biographical directory of mathematicians and scholars appears at the end of the book.)

In chapter 3, we present stories of solace in one's discipline as an escape from oppression and imprisonment. We start with Jean-Victor Poncelet, a captured officer in Napoleon's defeated army, who made great discoveries in geometry while a prisoner in Siberia, and we include José Luis Massera of Uruguay, who became a teacher to his fellow inmates in prison, giving them hope and determination to survive.

The potentially addictive nature of mathematics is examined in chapter 4. What drives a person to live only for mathematics? If that pursuit becomes an obsession, what are the possible psychic repercussions? We look at the story of Alexandre Grothendieck, one of the supreme mathematicians of the 20th century, who, after giving his life to his profession, became a hermit, repudiating his colleagues and the prevalent values of the 20th century. André Bloch engaged in disciplined daily mathematical work (and mastered univalent functions) in a psychiatric asylum at Charenton, France. The best known mad mathematician is Ted Kaczynski, "the Unabomber," whose psychopathology seems to be a murderous parody of mathematical reasoning. The tragic end of the great logician Kurt Gödel shows paranoia coexisting with genius.

Second on our list of four myths is the notion that *mathematics is a solitary pursuit*. This is refuted in chapters 5 and 6.

Intense and sustained thought requires a quiet environment and a highly focused mind-set. But discipline, discovery, and intellectual and emotional renewal thrive with support from caring connections. The popular notion of the solitary and eccentric mathematician is a distortion; it leaves out the rich social life of men and women in this field, including mentorship and collaboration.

In chapter 5, we explore friendships and partnerships in the lives of mathematicians. We look at the pleasures of thinking together that are vividly evoked by David Hilbert and Hermann Minkowski. We examine the complex collaborations of the eminent British mathematicians G. H. Hardy and John Littlewood, and the Indian mathematician Srinivasa Ramanujan. Their relationship is a powerful and bitter story of intellectual partnership and cultural isolation. We look at the professional and intimate connection between the bachelor professor Karl Weierstrass and the young Russian student Sonia Kovalevskaya, whose meeting gave rise to powerful emotional bonds and tragic loss. The singular mathematical marriage of Grace Chisholm and William Young presents surprising and shifting gender roles during their long partnership. The story of Chisholm and Young is particularly interesting, as mathematical marriages became quite frequent later in the 20th century.

In chapter 6, we examine the unique nature and culture of mathematical communities, and the support that they provide to both aspiring and established mathematicians. There are sometimes "golden" periods in the life of a cohesive group or department when inspired leadership, strong personal relationships, and institutional support result in a period of great productivity. This was achieved at Göttingen in the early 20th century. The legacy of Göttingen later inspired Richard Courant to establish a new mathematical community at New York University. (One of the authors, R.H., was lucky enough to spend several years there, first as a graduate student and later as a visiting postdoc.) Another famous mathematical community was the Bourbaki group in the French cities of Nancy and Paris. Forty years ago, Russian mathematicians experienced a "golden" decade of mutual encouragement and inspiration at the Department of Mechanics and Mathematics (Mekh-Mat) at Moscow University—and the impact of those years is still felt by the larger, international community. We also write of women who are struggling for equity, respect, and acceptance through the Association for Women in Mathematics, an organization that provides much needed support for its members.

The third of our four myths is often stated as a quote from G. H. Hardy, *that mathematics is a young man's game.* In chapter 7, we look at the issues of maturity, aging, and gender in the lives of mathematicians. We find that men and women continue to produce in their senior years, creating many ways to stay connected to their lifelong pursuits. We explore the seldom discussed experiences of mathematicians who are over 50—and in some cases even 70—years of age. This myth also assumes that mathematics is a male endeavor. It is refuted by the new reality of a steadily increasing number of women joining departments of mathematics and research institutions. While sexist attitudes linger and cause psychological damage, new female leadership has emerged. We recall some of the early female pioneers (Sophie Germain, Sonia Kovalevskaya, and Emmy Noether) and the accounts of contemporaries who discuss their commitment to a mathematical life (Mary Rudin, Joan Birman, Lenore Blum, Karen Uhlenbeck, et al.).

The last of the four myths is the notion that *mathematics is an effective filter for higher education.* Our last two chapters, in different ways, are about mathematics education, from the elementary up to the collegiate level. This topic cannot be avoided; mathematics education is a central part of mathematical life. There is an interesting paradox in how mathematics is perceived by the public at large. On the one hand, it is thought of as a world apart from practical human existence, and often it is taught as pure abstraction, which reinforces this perception. On the other hand, it is supposed to be the most useful qualification for students preparing for high-prestige professions such as engineering and architecture. In discussing this myth, we consider the role of mathematics in broader social contexts including elitist versus democratic perspectives on who should become a mathematician.

Some of the most emotional issues in the mathematical world are about the right and wrong ways to impart mathematics to children. In our last two chapters we confront these controversies frankly and argue for a realistic and humane approach to the difficult problem of improving math education in the United States.

In chapter 8, we review two opposite, yet intertwined, American educational traditions: the elite "Moore method" in Austin, Texas, and the universalist "Potsdam model" in upstate New York. These two narratives force us to face the issue of racial segregation in the history of mathematics in the United States.

This brings us in chapter 9 to the big question: How should mathematics be taught? Can we better serve those who say they like it? Must we disregard those who say they hate it? We address the never-ending call for improving mathematical achievement in the United States by offering some nonstandard, but realistic and humane, proposals that take into account the emotional aspects of learning.

In the book's conclusion, we envision a joyous and balanced view of the mathematical life of reason, emotion, and learning.

This is not a math education book of the kind that contains curriculum recommendations or statistical analyses of classroom experiments. But we do deliver the pain and pleasure of teaching, learning, and schooling. Neither is it the kind of math book that teaches some particular branch of mathematics. For example, when we refer to Fritz John's "functions of bounded mean oscillation" or Teiji Takagi's "class field theory," we only mention by name some of their mathematical contributions. For deeper exploration, one can turn to the plentiful literature of treatises and textbooks. However, we do attempt an overview of the most important work of Alexandre Grothendieck as part of understanding his emotional life story.

This book benefited from contributions to understanding mathematical life by nonmathematicians—psychologists, neuroscientists, anthropologists, and sociologists. We are grateful for these contributions and look forward to more such studies.

The writing of this book is a collaboration between a mathematician and a psychologist. Our disciplinary training and interests differ greatly, but we both view the life of the mind as a deeply humanistic activity. One eloquent spokesman for humanist philosophy was the physicist Jacob Bronowski. He wrote that the society of scientists must be governed by "independence and originality, dissent and freedom and tolerance: such are the first needs of science; and these are the values which of itself it demands and forms. The society of scientists must be a democracy. It can keep alive and grow only by a constant tension between dissent and respect; between independence from the views of others, and tolerance for them. The crux of the ethical problem is to fuse these, the private and the public needs."[4]

We hope that this joint examination will increase awareness of the richness of mathematical life in its many tensions between isolation and community, logic and playful exploration, and dissent and respect.

Bibliography

Bronowski, J. (1965). *Science and human values*. New York: Harper and Row.

Immordino-Yang, M. H., & Damasio, A. (2007). We feel, therefore we learn: The relevance of affective and social neuroscience to education. *Mind, Brain, and Education 1*(1), 3–10.

+1+

Mathematical Beginnings

A Passion for Numbers

How does a child first begin to become a mathematician? Is it a predisposition or some special gift? Does help and encouragement from parents contribute? What makes it possible, finally, to commit one's life to this risky, forbidding pursuit?

In this chapter, we tell contrasting stories about the childhood, adolescence, and schooling, up through graduate school, of some future mathematicians, both famous and not so famous. We also report the experiences of youngsters in Olympiad competitions, psychologists' investigations of prodigies, and what the parents of math prodigies are like.

A few famous mathematicians showed their interest and ability before school age. The Hungarian combinatorialist and number theorist Paul Erdős[1] (1913–1996) claimed that he independently invented negative numbers at age 4.

Stan Ulam (1909–1984) (sometimes referred to as "the father of the H-bomb") wrote in 1976: "When I was four, I remember jumping around on an oriental rug looking down at its intricate patterns. I remember my father's towering figure standing beside me and I noticed that he smiled. I felt, 'He smiles because he thinks I am childish, but I know these are curious patterns.'"[2]

Carl Friedrich Gauss (1777–1855), preeminent after Archimedes (c. 287 BC–c. 212 BC) and Newton (1643–1727), could calculate before he could read. In old age, Gauss told about this childhood feat: In a class that was told to add the numbers from 1 to 100, little Gauss produced the answer in a few seconds. (He constructed 50 pairs, grouping 1 with 100, 2 with 99, and so on. Each pair totals 101, and there are 50 pairs, giving the final sum, 50 × 101 = 5050.)

The famous physicist Freeman Dyson wrote in 2004:

One episode I remember vividly, I don't know how old I was; I only know that I was young enough to be put down for an afternoon nap in my crib. . . . I didn't feel like sleeping, so I spent the time calculating. I added one plus a half plus a quarter plus an eighth plus a sixteenth and so on, and I discovered that if you go on adding like this forever you end up with two. Then I tried adding one plus a third plus a ninth and so on, and discovered that if you go on adding like this forever you end up with one and a half. Then I tried one plus a quarter and so on, and ended up with one and a third. So I had discovered infinite series. I don't remember talking about this to anybody at the time. It was just a game I enjoyed playing.[3]

For some children, during times of personal upheaval, the simplicity and order of geometry and algebra are a comforting refuge. One example is the famous author and neurologist Oliver Sacks. During the bombardment of London in World War II, he was sent away from home and family. He writes:

For me, the refuge at first was in numbers. My father was a whiz at mental arithmetic, and I, too, even at the age of six, was quick with figures—and, more, in love with them. I liked numbers, because they were solid, invariant, they stood unmoved in a chaotic world. There was in numbers and their relations something absolute, certain, not to be questioned, beyond doubt. . . . I particularly loved prime numbers, the fact that they were indivisible, could not be broken down, were inalienably themselves. . . . Primes were the building blocks of all other numbers, and there must be, I felt, some meaning to them. Why did primes come when they did? Was there any pattern, any logic to their distribution? Was there any limit to them, or did they go on forever? I spent innumerable hours factoring, searching for primes, memorizing them. They afforded me many hours of absorbed solitary play, in which I needed no one.[4]

The childhood of the well-known mathematics educator Anneli Lax (1922–1999) was also disrupted by World War II. Mathematics was "the perfect sort of escape: I didn't have to look up anything; I didn't have to consult libraries or books. I could just sit there and figure things out."[5]

The Hungarian-born physicist Eugene Wigner (1902–1995) was diagnosed with tuberculosis at the age of 11 and spent weeks in a sanatorium in Austria. To help get through this difficult period he worked on geometry problems. "Sitting in my deck chair, I struggled to construct a triangle given only the lengths for the three altitudes. This is a very simple problem which I can do now in my dreams. But then it took me several months of concentrated effort to solve it."[6] Wigner attended a famous gymnasium in Budapest where John von Neumann (1903–1957) was also a student. They became lifelong friends. "It was likely the best high school in Hungary; it may have been the finest in the world."[7] "My heart was with numbers, not words. After a few years in the gymnasium I noticed what mathematicians call the Rule of Fifth Powers: That the fifth power of any one-digit number ends with that same number. Thus, 2 to the fifth power is 32, 3 to the fifth power is 243, and so on. At first I had no idea that this phenomenon was called the Rule of Fifth Powers; nor could I see why it should be true. But I saw that it was true, and I was enchanted."[8]

Steven Strogatz, an applied mathematician at Cornell, describes the fear and awe he felt when data he was plotting in a physics lab created a curve he had met in algebra class. He was recording how the length of a pendulum string affects the time for the pendulum to complete a swing. As he plotted the data on graph paper, he realized that

> these dots were falling on a particular curve that I recognized because I'd seen it in my algebra class—it was a parabola, the same shape that water makes coming out of a fountain. I remember experiencing an enveloping sensation of fear, then of awe. It was as if . . . this pendulum knew algebra! What was the connection between the parabolas in algebra class and the motion of this pendulum? There it was, on my graph paper. It was a moment that struck me, and was my first sense that the phrase "law of nature" really meant something. I suddenly knew what people were talking about when they said that there was order in the universe, and that, more to the point, you couldn't see it unless you knew math. It was an epiphany I've never really recovered from.[9]

We have limited evidence about childhood engagement in mathematics. However, the stories of those who have an early passion for numbers reveal their fascination with the patterns of mathematics. For others who are experiencing trauma, doing problems provides a refuge.

Mathematicians' Early Teen Years

Most famous mathematicians first showed a strong interest in mathematics around middle school age. For example, John Todd (1911–2007), a well-known international leader in numerical mathematics, said, "My mathematical career started in the following way. I was in a class, a singing class. My singing was so bad the teacher said I was disturbing the class and had to leave! There were some extra classes, ones with national examinations. And so I had to be put in one of them—it was a second-year algebra class! That's when I started learning mathematics."[10]

Another mathematician, Jenny Harrison, currently a professor at the University of California, Berkeley, was primarily interested in nature and music during her teen years. In an interview with John-Steiner, she recalled her beginnings. Jenny Harrison was born in Atlanta, Georgia, and spent most of her time outside of school in the woods. This attraction to the natural world may have influenced Harrison throughout her life and can be seen in the way she explores mathematical landscapes. She talked about her visual view of the world and her enjoyment in exploring paths in the woods. This strong preference may well have contributed to her later interests in geometry. The influence and encouragement of her older brother contributed to her strength and self-confidence. He got Harrison interested in basic problems in physics when they were teenagers.[11]

Although she clearly showed an aptitude for math early on (scoring the highest in her state in a take-home competition), her primary passion was for music. She studied the piano and believed for most of her adolescence that she would devote her life to it. Music is still a part of her life, but Harrison is basically a shy person and found that she was uncomfortable performing in public. "I knew I didn't want to do music and I went over and shut the lid to the piano. I was intrigued by three problems and wanted to try to understand them: the nature of consciousness, time and light. The question was how best to do this. I eventually settled on mathematics as I felt it would give me answers I could trust."[12]

Julia Robinson (1919–1985), who became famous for helping to prove that Hilbert's 10th problem is unsolvable (specifically, that there is no formula or computer program that can always decide whether an arbitrary polynomial equation with whole number coefficients has a whole-number solution) wrote, "One of my earliest memories is of arranging pebbles in the shadow of a giant saguaro, squinting because the sun was so bright. I think that I have always had a basic liking for natural numbers. To me they are the one real thing."[13]

Sophie Germain (1776–1831), who was to make an important contribution to proving Fermat's Last Theorem, was born in Paris. As a girl, she had to fight hard for the right to read mathematics. Her interest in mathematics began at age 13, during the French Revolution. Because of the battles going on in the streets of Paris, Sophie was confined to her home, where she spent a lot of time in her father's library. There she read about the death of Archimedes. It is said that on the day that his city, Syracuse, was being overrun by the Romans, Archimedes was engrossed in a geometric figure in the sand and ignored the questioning of a Roman soldier. As a result, he was speared to death.[14] If someone could be so engrossed in a problem as to ignore a soldier and die for it, thought Sophie, the subject must be interesting.

> Sophie began teaching herself mathematics, using the books in her father's library. Her parents did all they could to discourage her. She began studying at night to escape them, but they went to such measures as taking away her clothes once she was in bed and depriving her of heat and light to make her stay in her bed at night instead of studying. Sophie's parents' efforts failed. She would wrap herself in quilts and use candles she had hidden in order to study at night. Finally her parents realized that Sophie's passion for mathematics was "incurable," and they let her learn. Thus Sophie spent the years of the Reign of Terror studying differential calculus without the aid of a tutor![15]

Sofia Kovalevskaya (1850–1891) was the first woman to achieve full status as a professional mathematician. Born in Moscow in 1850, her interest in mathematics was stimulated in childhood by the wallpaper in her nursery. Kovalevskaya wrote, "When we moved to the country from Kaluga the whole house was painted and papered. The wallpaper had been ordered from St. Petersburg, but the quantity needed was not estimated quite accurately, so that paper was lacking for one room. Considering that all the other rooms were in order the nursery might well

Figure 1-1. Sofia Kovalevskaya, pioneer Russian analyst.
© Institut Mittag-Leffle

manage without special paper."[16] Some paper that was lying around in the attic was used for the purpose. By a happy chance the paper for this covering consisted of the lithographed calculus lectures of the analyst M. V. Ostrogradsky that her father had acquired as a young man. "It amused me to examine these sheets, yellowed by time, all speckled over with some kind of hieroglyphics whose meaning escaped me completely but which, I felt, must signify something very wise and interesting. And I would stand by the wall for hours, reading and rereading what was written there. I remember particularly that on the sheet of paper which happened to be in the most prominent place on the wall, there was an explanation of the concepts of infinitesimals and limits."[17]

Kovalevskaya recalled her uncle speaking to her about the quadrature of the circle: "If the meaning of his words was unintelligible to me, they struck my imagination and inspired me with a kind of veneration for mathematics, as for a superior, mysterious science, opening to its initiates a new and marvelous world inaccessible to the ordinary mortal."[18]

Kovalevskaya, too, had to overcome the resistance of her family. In fact, in order to study mathematics, she had to leave Russia for Germany. To do this legally, she had to marry, and so at age 18 she married "in name only," another idealistic student, Vladimir Kovalevskii.

The Uruguayan mathematician José Luis Massera (1915–2002), who in his fifties became an international cause célèbre as a political prisoner, described his discovery of mathematics as "a revolution." Massera wrote:

> When I was fifteen the revolution began in my home which lasted for several years. My father had a Hispanic American Encylopedic Dictionary of several volumes and of sufficiently high level. One day, on returning from German class, it occurred to me to look in the dictionary for one of the words that had been used, probably "equation." I found myself with an enormous quantity of different equations, whose existence I hadn't even suspected, nor how to approach them. My curiosity being amply satisfied with the algebraic equations, I went to look for one of the others of the dictionary. Thus, day by day and word by word, I began a review, entirely chaotic, no doubt, that was providing me with a harvest of mathematical terms and valuable information about them that I was accumulating and conceptualizing slowly.

During a family trip they passed through Paris where he accompanied his father to a great bookstore. There they found two books: one on classical geometry and the other on trigonometry. Massera devoured them in a short time. He liked the way the material was sequenced and practiced. Subsequently, with the help of a dictionary, he began to read a larger volume on projective geometry. He met his colleague and lifelong friend Rafael Laguardia in high school. He and Laguardia established a study group of young people interested in mathematics and shared their knowledge with them and each other.[19]

One of the most astounding childhood accomplishments in mathematics was achieved by Louis Joel Mordell (1888–1972), who was chair of the math department at the University of Manchester from 1922 to 1945, and then successor to G. H. Hardy in the Sadleirian chair at Cambridge University. This pillar of English pure mathematics was a self-taught boy from Philadelphia. He was the third of eight children of a Hebrew scholar. Louis became fascinated by mathematics while in grade school. He learned on his own, buying second-hand mathematics books from a bookstore in Philadelphia for 5 or 10 cents when he was 13 years old. These books contained problems from the Cambridge tripos examinations (written tests for undergraduates), so Louis came to think of Cambridge University in England as the mathematical summit. He later wrote, "I conceived what can only be described as a thoroughly mad and crazy idea of going to Cambridge and trying for a scholarship. . . . I had no idea of the necessary standards. I was self-taught mathematically and had never participated in a competitive examination."[20] He earned the money for his passage to England by tutoring his fellow pupils for 7 hours a day. In December 1906 he went to Cambridge and competed in the university scholarship exam. He placed first! He could only afford a one-word telegram back to his father. It read simply, "Hurrah."

The Russian, Israel Moiseevich Gelfand (1913–2009), is one of the most illustrious mathematicians of the 20th century. In the small town near Odessa where he grew up, there was only one school. The mathematics teacher there was very kind, but Israel Moiseyevich quickly surpassed him. Gelfand said, "My parents could not order books in mathematics for me; they had no money. But I became lucky. When I was 15 my parents brought me to Odessa for an appendectomy. I said that I wouldn't go to the hospital unless they bought me a book on

Figure 1-2. Israel M. Gelfand, one of the foremost
mathematicians of his time. © Mariana Cook 1990.

mathematics."[21] Before he read that book, he had thought that
algebra and geometry were two different mathematical subjects.
When he saw Maclaurin's formula for the sine, he suddenly realized that there was no gap between the two. "Mathematics
became united. And since then I understood that different areas
of mathematics together with mathematical physics formed a
single whole."[22]

Andrew Wiles, who is now famous for proving Fermat's Last
Theorem, first became fascinated by that problem at the age of
10. He loved doing math problems in school and would take
them home and make up new ones of his own.[23] When he was
10, Andrew found Fermat's Last Theorem in Eric Temple Bell's

classic *Men of Mathematics*. Thirty years later, Wiles remembered, "It looked so simple, and yet all the great mathematicians in history couldn't solve it. Here was a problem that I, a ten-year-old, could understand and I knew from that moment that I would never let it go. I had to solve it."[24]

Psychological Themes

David Feldman, a developmental psychologist at Tufts University, studied child prodigies and didn't find many prodigies under the age of 10. He wrote, "Mathematics fosters far fewer prodigies than I thought would be the case when this work began."[25] With the help of Julian Stanley at Johns Hopkins University, he did identify Billy Devlin. When Devlin was 6 he scored in the high school range on the Scholastic Aptitude Test. He became one of the participants in Feldman's study at age 7. But as he grew older, he turned away from mathematics to physics and astronomy.

Recently, Terence Tao received international attention as a 2006 Fields Medal winner. "Terry is like Mozart; mathematics just flows out of him," said John Garnett, professor and former chair of mathematics at the University of California, Los Angeles, "except without Mozart's personality problems; everyone likes him. Mathematicians with Terry's talent appear only once in a generation. . . . He's an incredible talent, and probably the best mathematician in the world right now."[26] He is a prodigy who started to play with numbers at 2 years of age. His father, Billy Tao, is a pediatrician who, together with his mother, carefully supported their child's great gift. They were successful in setting up an individualized program for him where he was able to acquire each subject at his own pace, quickly accelerating through several grades of math and science while remaining closer to his age group in other subjects.[27] His home environment was supportive, and today he continues this legacy by being a thoughtful and devoted parent. At the same time, he produces an extraordinary range of important mathematics. He is a wonderful counterexample to the popular image of the eccentric mathematician.

In the psychologist Ellen Winner's comprehensive work on gifted children, she writes of Ky Lee, a toddler who loved letters and numbers, starting at age 18 months. By age 2, his favorite toys were plastic numbers and blocks with numbers on them. He said the numbers over and over as he handled the objects. When he was 3, he was on a camping trip with his parents. When the ranger at the park gate asked for their license number, neither parent remembered it, but Ky Lee answered with ease, "502-VFA."[28] Ky Lee also could do mental addition, a feat shared in childhood by a few famous mathematicians.

Another recent prodigy was Ganesh Sittampalam, who revealed exceptional insight into subtraction at age 5. He was coached by his father, who had a Ph.D. in mathematics. His father said, "I felt it was very important that he didn't learn anything by rote. He had to understand the conceptual and logical structure behind the whole thing. I made sure he thought for himself and I always stopped short of telling him what the next step was."[29]

Some researchers suggest that ease with calculations comes from living with numbers, being fascinated by them, and representing them as a mathematical landscape—being on familiar terms with numbers and knowing them inside out.

In the most comprehensive study done so far of mathematicians raised in the United States, William Gustin interviewed 20 mathematicians who had won a Sloan Foundation Fellowship. The majority of their fathers had advanced degrees, and their mothers were also well educated. The parents had gone to college during the Depression, and they had a deep commitment to learning. The value of intellectual and educational achievement was transmitted to the children. Working hard, doing well, and being precise were values that their parents taught them.[30]

Steve Olsen studied U. S. participants in the 2001 International Mathematical Olympiad. The participants' parents reported their children's early interest in puzzles and Legos. One mother remembered that her son, Tiankai Liu, had been fascinated with the size and shape of manhole covers. Spatial visualization and an interest in patterns are two of the most often noted interests of future mathematicians. Gustin quotes the father of one Sloan award winner, "He would spend hours building a tower of blocks, precariously balanced. There would be a wail of exasperation and anguish when it finally collapsed. And then he would start redoing it."[31]

The parents of gifted children are often described as highly attentive, actively stimulating and teaching their children. The families are usually stable and harmonious, with great warmth and nurturance. Independence and autonomy are encouraged.[32] Young people raised in such families work up to their potential more often than those from less supportive homes, and they are more independent and original.[33] Many of the mathematically gifted children in the United States (or their parents) are immigrants from other cultures. Margaret Murray found that one-third of the women mathematicians who received Ph.D.'s in the United States in the 1940s and 1950s were immigrants, or children of immigrants, from eastern and central Europe. These immigrant families brought with them the respect for learning and culture that was the hallmark of many European societies and strongly emphasized in the Jewish tradition. In many of these families, the dream of a better life in a land of opportunity was an extremely powerful motivating force for both parents and children, encouraging them to achieve in work and at school.[34] Some of the women Murray studied were daughters of highly educated professionals. Others had at least one parent who had attended some college.

More recently, a high number of young participants in science and mathematics competitions are from East Asian families, either native-born citizens or immigrants. A study by Stuart Anderson of the 2004 Intel Science Search found that 7 of the top 10 award winners in the year's contest were immigrants or their children. Of the top 40 finalists, 60 percent were the children of immigrants.[35] One immigrant child, Tiankai Lu, explained one reason why he chose mathematics during his early schooling: "I wasn't super-duper in English, partly because my parents didn't know English very well . . . so I decided that maybe I could do math."[36]

Some parents spend many hours tutoring and encouraging their children in mathematics. Two who succeeded were Leo Wiener, the father of Norbert Wiener, and Tobias Dantzig, the father of George Dantzig (1914–2005), the inventor of the simplex method of linear programming. Leo Wiener was a self-educated scholar who became a professor of Slavic languages at Harvard. In the July 1911 issue of *American Magazine*, his strong belief in early training as the source of precocious mental development is described.[37] The article reports that Leo Wiener followed his strong convictions by making his children the subjects of an education experiment. He said to the reporter, "It is all nonsense to say that Norbert (and Norbert's sisters Constance and Bertha) are unusually gifted children. They are nothing of the sort. If they know more than other children of their age, it is because they have been trained differently." In Norbert's autobiography, *Ex-Prodigy*, he wrote of his father's algebra instruction. "He would begin the discussion in an easy, conversational tone. This lasted exactly until I made the first mathematical mistake. Then the gentle and loving father was replaced by the avenger of blood. . . . The very tone of my father's voice was calculated to bring me to a high pitch of emotion. . . . My lessons often ended in a family scene. Father was raging. I was weeping and my mother did her best to defend me, although hers was a losing battle."[38] Even while Norbert was a student at Tufts College, Leo continued to monitor his son's homework. It required putting the Atlantic Ocean between them to emancipate the son from the father. Wiener's student Norman Levinson wrote of his teacher, "[E]ven forty years later when he became depressed and would reminisce about this period, his eyes would fill with tears as he described his feelings of humiliation as he recited his lessons before his exacting father.

Fortunately he also saw his father as a very lovable man and he was aware of how much like his father he himself was."[39]

George Dantzig's story is happier. His father, Tobias, was a well-known mathematician who had studied with Henri Poincaré (1854–1912) in Paris and who wrote a book, *Number, the Language of Science*, that is still one of the best popularizations of advanced mathematics. He provided strong support to George, who wrote, "He gave me thousands of geometry problems while I was still in high school . . . the mental exercise required to solve them was the great gift from my father. Solving thousands of problems during my high school days—at the time when my brain was growing—did more than anything else to develop my analytical power."[40]

Personal Characteristics

Gustin found that both the participants in his study and their parents had the ability to devote long periods of time to a single activity.[41] In Steve Olson's book, *Count Down*, the stories of successful Olympiad competitors revealed an extraordinary capacity for mental focus; they were willing to stay with a problem for hours, or even days. If there is one quality that young mathematicians share, it is this power of concentration.

Feldman writes of prodigies' intense dedication to their field, their extreme self-confidence, and their mixture of adult- and childlike qualities.[42] Recalling their childhoods, many creative individuals remember persistence, enthusiasm, energy, and determination. Some of them say they had a great thirst for knowledge, an exceptional sense of direction, or even an obsession with the problems they pursued. "They often have an unusual ability to resist the distractions of everyday life, to ignore discouragement or ridicule, or to persist in working toward their goals in the face of repeated failures."[43]

The most consistently mentioned quality of gifted young mathematicians is curiosity. One parent recalls, "He asked questions at a very young age—constantly asked questions. He just couldn't wait to learn."[44] There are similar reports from almost all the parents, and what makes parents of mathematicians unique is their response to their children's questioning. They respond seriously, often encouraging even more questions.[45] Billy Devlin, the math prodigy studied by David Feldman, had a voracious appetite for information and made astonishing progress in his work with a math tutor.[46] He had a passion for collecting things and arranging them in order. He knew a lot about natural science, science fiction, geography, arithmetic, and mathematics.[47, 48]

Between 1955 and 1956, Krutetskii compared mathematically gifted Russian school children with their peers. He found, as many others have reported, that students who excel in mathematics spend a lot of time thinking about the subject. He emphasizes the students' flexibility in thinking. While they know how to use a previously practiced solution, they can readjust it when the practiced solution fails. Like the Olympiad contestants and other successful young mathematicians, they look for direct and elegant solutions. They tire less in math class than in more verbal classes. Ellen Winner distinguishes between "globally gifted" children and those specifically gifted in mathematics. One globally gifted child mastered reading when he was 3 years old and showed as much interest in numbers as in words. He would ask his parents to give him addition and subtraction problems to do in his head. He could mentally add two-digit numbers if there was no carrying involved. Because he usually solved math problems in his head, he had difficulty in school when his teacher insisted he write out math problems with conventional symbols.[49]

Winner describes one gifted child who likes being different and doesn't mind being alone a lot of the time. This is a typical story. Gifted youth spend more time alone than average youth. Like most people, they feel happier when with others, but they do not mind solitude as much as most.[50] Half of the mathematicians in Gustin's study seem to have handled isolation and long periods of solitary study quite well. But other mathematicians found being "different" difficult. Their social needs were met only after they had the chance to take college math courses or to meet other young people interested in mathematics. Students like these are helped by summer programs for gifted children, where they find peers who share their passions and perseverance.

Many gifted children do have normal social lives, but still they cherish their independence. Their self-confidence and desire to be in control of their own activities can make schooling a challenge. They like to choose books themselves and to focus on topics not taught in class. They enjoy taking college classes while they are still in high school, and they prefer to have a flexible curriculum.

Radford quotes authors who find that young mathematicians have a passion for counting, and they often include numbers in stories and rhymes. They like to use logical connectives (if, then, so, because, either, and or). They delight in making patterns showing balance and symmetry. They like jigsaw puzzles and construction toys, and they arrange their toys neatly and precisely. They use sophisticated criteria in sorting and classifying.[51]

The young mathematicians Olson wrote about found it important to picture mathematical problems visually. Some famous physicists have possessed this ability. In an oft-quoted account, Einstein said that he first thought about special relativity by imagining he was riding on a wave and watching the wave behind him.

Some mathematicians in Gustin's study were interested in the way things work. They liked to take toys apart and look at the gears, valves, gauges, and dials.[52] Half of the mathematicians in the study were interested in scientific and mechanical projects, and in building models, before the age of 12. "I think I spent a lot of time by myself as a youngster. The first dollar I ever saved I spent on a model airplane. I sanded it, glued it and put it all together and painted it. I just fell in love with the whole process."[53]

Teachers

Many mathematicians' interest in math was stimulated by a teacher. The exceptional instruction at the Lutheran Gymnasium in Budapest, Hungary, produced aerodynamicist Theodor von Kármán (1881–1963), mathematician John von Neumann (1903–1957), and quantum physicist Eugene Wigner (1902–1995). All became world-famous and major contributors to U.S. science. Wigner fondly remembered his mathematics teacher, László Rátz, who "loved teaching; he knew his subject and how to kindle interest in it. He imparted the deepest understanding. Many . . . teachers had great skill but no one could invoke the beauty of the subject like Rátz."[54]

On the other hand, some talented girls were not supported in the study of mathematics by their teachers. When Alice Schaefer was skipped from third to fourth grade, her teacher said that although she and one of her classmates were expected to do all right in fourth grade, the teacher doubted that the two girls would be able to do long division. Alice was indignant. She subsequently recounted that she was determined to learn long division in the fourth grade. The experience gave rise to her first feelings about mathematics.[55] Then her high school math teacher opposed her continuing in math even though she excelled in his class. When Schaefer asked him to recommend her to the University of Richmond, he answered, "If you want to major in mathematics I won't write for you because girls can't do mathematics."[56] But Schaefer prevailed. Other female students were more fortunate. For instance, the Canadian mathematician Margaret Marchand was encouraged by her teacher, Mr. Robson. He recognized her aptitude and was the first person to suggest that she go on to the university.

Some women did not choose mathematics until college. Growing up in the 1950s, Judith Roitman never imagined herself as a mathematician. She started writing poetry at age 8 and was also a minor music prodigy, but as a girl she was discouraged from a career in mathematics. "Roitman remembers feeling that no matter how well she did in mathematics she could never achieve real understanding, because real understanding was, by definition, something only boys could have."[57]

Planning to become a high school English teacher, Roitman went to Sarah Lawrence, an all-women's college at the time. There she changed her beliefs about what she could and couldn't do. First, she studied poetry and language. Then she moved to philosophy. Finally, she turned to mathematics because the fact that mathematicians were constantly inventing new ways of thinking appealed to her.[58] But in the late 1960s and early 1970s gender bias followed her through graduate school. The climate was a challenge for women in mathematics. Despite such an environment, Roitman succeeded, with the support of a peer group of female mathematicians, graduate students, and postdoctoral instructors, and with Mary Ellen Rudin as her mentor. (Rudin is a leading topologist. We quote extensively from her in chapter 9.) Rudin provided Roitman with an existence proof, that is, an example of a woman with a salaried professional life as a research mathematician.[59] Roitman became a leading researcher in set theory and a professor at the University of Kansas.

In her study of how women became mathematicians, Margaret Murray (2000) found that nearly all the women she interviewed had at least one college teacher who encouraged and influenced them, even at a time when women in mathematics challenged the prevalent social norms. Pregnant women were discouraged from attending class. It took a committed teacher to encourage a pregnant student to persist in her studies. Students in women's colleges benefited from the mentorship of female professors. At Bryn Mawr College, the lasting influence of Emmy Noether (1882–1935), one of the most distinguished mathematicians of the early 20th century, lasted even after her death. (Noether was one of the creators of modern abstract algebra.) One Bryn Mawr graduate recalled, "They were still very much under the spell of having Emmy Noether there. There were still stories going on about her."[60]

These accounts and studies can assist us in regard to how mathematical talent originates and develops. It appears unexpectedly. It is not created at will by parents or teachers, but their support is crucial for developing it. Intense interest in mathematics does not usually manifest itself until the age of 10 or 12. Often a future famous mathematician does not find his or her vocation until their late teens, or even later. It seems that some developing inner inclination or aptitude becomes visible only when a certain intellectual maturity has been achieved and a supporting environment has been provided.

Contests and Competitions

Melanie Wood was the tutor for the U.S. Mathematical Olympiad team in 2001. She first became interested in mathematics in kindergarten, and remembers that she got in trouble for it. Once she was playing with flash cards that had numbers on them, and she sorted them into evens and odds. "I was realizing things like when you add two odd numbers, no matter which two they were, you always got an even number. And when you added an even number to an odd number—things like that. I got in trouble because I wasn't supposed to be playing with those flash cards. I had already passed that level and I was supposed to be playing with some other flash cards."[61]

Wood's mother played math games with her. When Melanie was very young, her mother did not think that she was particularly outstanding in mathematics. But by 7th grade Melanie was in an accelerated math class, although math was only one of her many interests. That year she was invited to take part in a national competition called Mathcounts. Participants compete at first in individual schools and then advance to regional contests. The four top individual scorers in each state go on to the national competition. In the 7th grade Melanie Wood had little advanced preparation for Mathcounts, but she finished first in Indiana. "That really flipped my world around. In terms of

Figure 1-3. Melanie Wood, Olympiad contestant and coach, mathematical researcher. Courtesy of Archives of the Mathematisches Forschungsinstitut Oberwolfach.

making math something important in my life, and changing my view of who I was and what I was good at."[62]

When Melanie entered high school, she was asked to train for the International Mathematical Olympiad. At the training camp she was the only girl. This was an emotional challenge to her, which she overcame with the help of a Bulgarian woman who was teaching mathematics at Harvard. "Having a role model like that was a big deal in my life. Previously I never knew a mathematician that I could look at and think, in 10 years I want to be like that person. And so it was hard to imagine becoming a mathematician."[63] In subsequent years she attained many honors and is now a Szegö assistant professor at Stanford University.

Problem competitions arouse great excitement among young mathematicians. Gábor Szegö (1895–1985), who ultimately became chair of the mathematics department at Stanford University, remembered his excitement as a student in the years from 1910 to 1912 at the monthly arrival of the Hungarian high school math newspaper. "I would wait eagerly for the arrival of the monthly issue and my first concern was to look at the problem section, almost breathlessly, and to start grappling with the problems without delay. The names of the others who were in the same business were quickly known to me, and frequently I read with considerable envy how they had succeeded with some problems which I could not handle with complete success, or how they had found a better solution (that is, simpler, more elegant or wittier) than the one I had sent in."[64]

In the United States, many promising students also thrive on math competitions and the Intel Science Talent Search (formerly the Westinghouse Science Talent Search). This is in contrast with their high school experiences, where many reported that their teachers were not ready to meet the challenge of a highly motivated student. Some of the young mathematicians receive individualized tutoring or attend university classes at an early age. Sixteen of the 20 mathematicians in Gustin's study did independent work in mathematics while they were in high school. They read books that their parents or older siblings had used in college, and some read scientific magazines. But they found few opportunities to talk mathematics with gifted peers in high school. This is why summer math camps were established in the 1950s. These were more challenging and interesting than regular class. There they explored fascinating new topics and developed their own techniques for solving problems. They discovered the excitement of doing something well and being recognized for it.[65]

Mentors in Graduate School

The most critical socializing influence on future mathematicians is graduate school. A mathematics department's reputation is determined mainly by the quality of its research. A student

Figure 1-4. Norbert Wiener. Courtesy of Smithsonian Institution Libraries, Washington, DC

goes to a top-ranked department hoping to meet outstanding professors and fellow students deeply involved in mathematics. A graduate school class in mathematics lets the student see a mathematician at work.

In graduate school it is possible for students to establish personal relationships with teacher mentors. A powerful example of such a relationship was that of Norman Levinson (1912–1975) and his mentor Norbert Wiener (1894–1964). Levinson writes, "I became acquainted with Wiener in September 1933, while still a student of electrical engineering, when I enrolled in his graduate course. It was at that time really a seminar course. At that level he was a most stimulating teacher. He would actually carry on his research at the blackboard."[66] As soon as Levinson showed some slight comprehension of what Wiener was talking about, Wiener handed him the manuscript of a treatise he was writing. Levinson found a gap in one of Wiener's arguments and supplied the missing reasoning. "Wiener thereupon sat down at his typewriter, typed my lemma, affixed my name and sent it off to a journal. A prominent professor does not often act as secretary for a young student. He convinced me to change my course from electrical engineering to mathematics. He then went to visit my parents, unschooled immigrant working people living in a run-down ghetto community, to assure them about my future in mathematics. He came to see them a number of times during the next five years to reassure them until they finally found a permanent position for me. (In these depression years positions were very scarce.)"

The impact of some particularly gifted teachers goes beyond the subject matter—it communicates the passion of those with experience to the novice. In recollecting his early years, Herbert Robbins (1915–2001)—a famous statistician and coauthor with Richard Courant of *What is Mathematics?* said of his Harvard professor, Marston Morse:

> There was something going on in his mind of a totally different nature from anything I had seen before. That's what appealed to me. . . . He was a father figure to me—my own father died when I was 13. Marston and I were about as different as two people could be, we disagreed on practically everything. And yet, there was something that attracted me to Marston that transcended everything I knew. I suppose it was his creative, driving impulse—this feeling that your house could be on fire, but if there was something you had to complete, then you had to keep at it no matter what.[67]

(Morse theory studies the properties of gradient vector fields. It connects with many parts of pure and applied mathematics.)

Stan Ulam's (1909–1984) career was stimulated by his rapport with the set theorist Kazimir Kuratowski in the Polish city of Lvov:

> From the very first lecture I was enchanted by the clarity, logic, and polish of his exposition and the material he presented. From the beginning I participated more actively than most of the older

Figure 1-5. Richard Courant and Herbert Robbins. Mathematical People: Profiles and Interviews. Eds. Donald J. Albers and G. L. Alexanderson. Boston: Birkhauser, 1985. Pg. 285. With kind permission from Springer Science and Business Media.

students in discussion with Kuratowksi. . . . I think he quickly noticed that I was one of the better students; after class he would give me individual attention. . . . Soon I could answer some of the more difficult questions in the set theory course, and I began to pose other problems. Right from the start I appreciated Kuratowski's patience and generosity in spending so much time with a novice. Several times a week I would accompany him to his apartment at lunchtime, a walk of about twenty minutes, during which I asked innumerable questions. Years later, Kuratowski told me that the questions were sometimes significant, often original, and interesting to him. . . . At that time I was perhaps more eager than at any other time in my life to do mathematics to the exclusion of almost any other activity.[68]

Sometimes a chance encounter is a catalyst for a new under-standing of one's field. Paul Halmos (1916–2006) described having lunch with the famous probabilist Joe Doob at a drug-store in Urbana, Illinois. "My eyes were opened. I was inspired. He showed me a kind of mathematics, a way to talk mathe-matics, a way to think about mathematics that wasn't visible to me before. With great trepidation, I approached my Ph.D. supervisor and asked to switch to Joe Doob, and I was off and running."[69]

We have already mentioned the father and son Tobias and George Dantzig. After reading papers by the famous statistician Jerzy Neyman (1894–1981), George went to study in Neyman's department at Berkeley.

> During my first year at Berkeley I arrived late one day to one of Neyman's classes. On the blackboard were two problems which I assumed had been assigned for homework. I copied them down. A few days later I apologized to Neyman for taking so long to do the homework—the problems seemed to be a little harder to do than usual. I asked him if he still wanted the work. He told me to throw it on his desk. I did so reluctantly because his desk was covered with such a heap of papers that I feared my homework would be lost there forever. . . . About six weeks later, one Sunday morning about eight o'clock, Anne and I were awakened by someone banging on our front door. It was Neyman. He rushed in with papers in hand, all excited: "I've just written an introduction to one of your papers. Read it so I can send it out right away for publication." For a minute I had no idea what he was talking about. To make a long story short, the problems on the blackboard which I had solved thinking they were homework were in fact two famous unsolved problems in statistics. That was the first inkling I had that there was anything special about them.[70]

Dantzig became one of the major statisticians in the United States and received the National Medal of Science.

In Budapest, in the years before and after World War I, Lipot Fejér (1880–1959) was the mentor to a whole generation of Hungarian mathematicians. He had been a great problem solver in high school. László Rátz often opened his problem session at the Lutheran Gymnasium in Budapest with the announcement, "Lipot Weisz has again sent in a beautiful solution." Weisz, who later changed his name to Fejér, won second prize in the

Eötvös competition. (That competition is the grandfather of Mathcounts and the U.S. Mathematical Olympiad.)

Fejér became a professor in Budapest in 1911. George Polya wrote, "Almost everybody of my age group was attracted to mathematics by Fejér. Fejér would sit in a Budapest cafe with his students and solve interesting problems in mathematics and tell them stories about mathematicians he had known. A whole culture developed around this man. His lectures were considered the experience of a lifetime, but his influence outside the classroom was even more significant."[71]

One of Fejér's students, Agnes Berger, recalled that he gave very short, very beautiful lectures that lasted less than an hour. "You sat there for a long time before he came. When he came in, he would be in a sort of frenzy. He was very ugly-looking when you first examined him, but he had a very lively face with a lot of expression and grimaces. The lecture was thought out in very great detail, with a dramatic denouement. It was a show."[72] Fejér was a major contributor to Fourier analysis (the expansion of general functions in series of sines and cosines).

Another quality of good mentors is that they understand the tensions graduate students confront—the need to balance discipline and commitment as a learner with the complex responsibilities of adulthood. The Berkeley professor S. S. Chern (1911–2004) was able to respect and work with the challenges his students faced. Following are two tributes from the book, *S. S. Chern: A Great Geometer of the Twentieth Century.*[73]

Chern had agreed to take on Louis Auslander as a student at the University of Chicago before Auslander took his Ph.D. qualifying exam in the spring of 1952. But the night before the geometry part of the exam, Auslander's wife arrived home with a newborn son, and hemorrhaged. The next morning he performed poorly in geometry. But when Auslander saw Chern and asked him if he would still be his thesis adviser, Chern

conveyed the understanding that examinations were not important—it was now the time to do mathematics. Then began

Figure 1-6. Shiing-Shen Chern. Mathematical People:
Profiles and Interviews. Eds. Donald J. Albers and
G. L. Alexanderson. Boston: Birkhauser, 1985. Pg. 285.
With kind permission from Springer Science and
Business Media.

a process of education, an apprenticeship, by indirection. Chern
would say things like "Would you look at Finsler geometry?"
or, "It would be very nice if we meet in my office one day a
week and talk things over." No matter what I presented, Chern
would listen politely and almost silently. On occasion he would
say, "I do not understand." I soon learned that "I do not understand" was a euphemism for "That's wrong!!" Somehow Chern
conveyed the philosophy that making mistakes was normal and
that passing from mistake to mistake to truth was the doing of
mathematics. And somehow he also conveyed the understanding that once one began doing mathematics it would naturally
flow on and on. Doing mathematics would become like a stream
pushing one on and on. If one was a mathematician, one lived
mathematics . . . and so it has turned out.[74]

Philip A. Griffiths, former director of the Institute for Advanced Study at Princeton stated:

> Chern is genuinely interested in the work and ideas of students just finding their way. He is encouraging yet is willing to say some idea may not be interesting. He demonstrates a combination of wisdom, mathematical discrimination and tact. He always treats one with respect, as a colleague and equal. In addition to the mathematical relationship, he shows a real interest in the person in a broader sense, asking about his family, career plans, and travel, discussing world politics, history and events with as much wisdom as he shows in mathematical discussions. Long before the concept of "mentoring" came into vogue Chern was a model mentor. For those just embarking on a career in mathematics, as I was those thirty-odd years ago, the experience described above can be decisive. A beginning student needs to learn more than facts and techniques: he or she needs to absorb a world view of mathematics, a set of criteria with which to judge whether or not a problem is interesting, a method of passing on mathematical knowledge and enthusiasm and taste to others. To most fully develop as a mathematician one needs a mentor who can provide what Chern has provided for so many: formal teaching, teaching by example, encouragement, realism, and contacts.[75]

While male mathematicians freely give credit to a mentor in informal conversations, we have found that women publish such acknowledgments more often. At the Courant Institute at New York University (NYU), Lipman Bers (1914–1993), a leading researcher on elliptic partial differential equations and Riemann surfaces, had a strong commitment to women. During the 1950s he helped Tilla Weinstein (later Tilla Klotz Milnor) continue her studies when she became pregnant, while others, including the dean, resisted her efforts. Bers continued to advise and support his students after they left the caring environment of NYU.

One Courant Institute student, the group theorist Rebekka Struik, so much enjoyed her close contact with Wilhelm Magnus, another well-known professor there, that she was disappointed when Magnus told her that her work was complete and that it was time for her to look for a position.[76] Her father was Dirk Struik, a math professor at Massachusetts Institute of Technology (MIT), an authority on differential geometry, and a Marxist. He was one of the "Massachusetts Ten" indicted in the 1950s on the unusual charge of conspiracy to overthrow the Commonwealth of Massachusetts. The case against the Massachusetts Ten was thrown out on appeal, and Dirk Struik was restored to his teaching position at MIT.

Richard Courant (1888–1972) himself was extraordinarily supportive of women and helped them work out the conflict between their home responsibilities and their studies. There was another female student at the Courant Institute whose father was a well-known mathematician. Cathleen Morawetz is the daughter of J. L. Synge, a leading applied mathematician from Ireland who was long a professor at Toronto. (Her great-uncle was the famous Irish playwright of the same name.) Courant and Synge met at a math meeting and found that the daughters of both of them had recently married. The two distinguished fathers bemoaned the likelihood that their daughters wouldn't go on in their chosen fields, mathematics and biology, respectively. "Ja, ja, well, you can't do anything about my daughter," Courant sighed, "but maybe I can do something about yours. You should send her to see me some time."[77] Morawetz did become a student at the Courant Institute, earned a Ph.D. there with a thesis on transonic flow written under K. O. Friedrichs, became a professor there, ultimately became head of the institute, was elected president of the American Mathematical Society, and subsequently became a recipient of the National Medal of Science.

When Hersh started graduate work at Courant, he was enrolled in Morawetz's course, Introduction to Applied Mathematics, and was hired to grade papers in that course. A student grading papers for a class he attends would be considered irregular in most university departments, but it was no problem at Courant, which was run as a family enterprise. (It was joked there that "nepotism is compulsory.") Both of Courant's daughters married mathematicians—Jerry Berkowitz was a long-time professor at the Courant Institute, and Jürgen Moser, famous for work in dynamical systems, spent many years at the Swiss Federal Institute of Technology (ETH) in Zurich. There is an old (somewhat sexist) saying, "Mathematical talent is inherited from the father-in-law to the son-in-law." In fact, Courant, with only two daughters, managed to have three mathematician sons-in-law! (Some years after Jerry Berkowitz passed away, Lori Berkowitz, nee Courant, married Peter Lax—another important member of the Courant Institute.) The day is surely coming when we will give examples of mathematicians as mothers-in-law and daughters-in-law.

Many mathematicians have said that their greatest satisfaction as professors has been working with thesis students. Nurturing and developing a mathematical mind from small beginnings to full-grown power can be wonderfully fulfilling. When a student you have nurtured comes to you with a fruitful new idea about a problem you have long struggled with, the pleasure of such a collaboration can be even greater than the pleasure of solving a problem on your own. One famous example of such a relation was between Karl Weierstrass (1815–1897) and Sofia Kovalevskaya (1850–1891) about which we write further in chapter 5.

On the other hand, not all relationships between thesis student and thesis supervisor are rosy and cheery. The relation can be close and intimate or it can be distant and formal. One famous mathematician described supervising a dissertation as "research by the professor, done under difficult conditions." The supervisor of one mathematician we know did absolutely nothing to help him further once he had finished his dissertation. Another told us that he dreaded meeting with his supervisor, who would lose his temper and throw chalk during their meetings. Agnes

Berger, whom we quoted above in describing Fejér's lectures in Hungary, said: "I was greatly amazed when I saw that in America a professor would sit down with a graduate student. Nothing like that ever happened in Budapest. You would say to the professor, 'I'm interested in this or that.' And then eventually you would come back and show him what you did. There was none of the hand-holding that goes on here. I know people here who see their students every week! Have you ever heard of such a thing?"[78]

Students receive guidance and inspiration not only face to face but also at a distance. George David Birkhoff (1884–1944), the dominant American mathematician of the early 20th century, received his inspiration from Henri Poincaré of Paris.[79] Learning from "distant teachers,"[80] whose impact is through their published work rather than from face-to-face interaction, is a critical aspect of creative development. When creative individuals discover their own teachers from the past, there can be a recognition of an intense, personal kinship, as the work of another invokes a special resonance in them. Once such a bond is established, the learner explores those valued works with an absorption that is the hallmark of creative individuals. In this way, they stretch, deepen, and refresh their craft and nourish their intelligence, not only during their early years of apprenticeship but repeatedly through the many cycles of their work lives.[81]

For a student first embarking on a mathematical life, the work of one's predecessors is an intimidating challenge. The great achievements in the field seem overpowering. Often, when facing self-doubt or discouragement, these men and women lean on the support of their mentors.

Preparing for the Mathematical Life

For students of mathematics it is an intense experience to commit oneself to this discipline, where there is no guarantee of success. Research in mathematics requires deep focusing on a problem for long periods of time. One's previous accomplishments by no means guarantee success at a new problem. Some participants in Gustin's study still felt insecure about their ability to do original mathematics, even after receiving a Ph.D. "I really had grave doubts about my ability to do creative mathematics. There is no way of knowing until you do it."[82] Sometimes a graduate student chooses a thesis topic only to find out that although initially it was promising, ultimately it becomes a dead end.

What drives such a concentrated effort when the end result is not predictable? "You have to immerse yourself in something. Think about it constantly. You work and work and get these ideas floating around and you have to reach a certain threshold and then some of the problems solve themselves. Some of them come to you two years later, but you have to concentrate, to focus."[83] And even after you succeed, recognition comes only from a small group of specialists.

But the excitement of obtaining a new result can be deeply satisfying. "I know for the best two or three things that I have ever done, there is a feeling of awe that is just incomparable. I feel privileged to have added a couple of things to the field. I love the subject, there is nothing I enjoy more than finding the solution to a problem after a lot of work. And even though it is occasionally very painful to fail, after working on some problem for a year or two, there is still in the back of your mind the thrill of the chase."[84]

The choice of becoming a mathematician presents the young person with joy and fear. In this chapter we have sketched the conditions that support an individual's ability to take enormous risks in one's intellectual life. Parental response to early curiosity, inspired teaching, and lively interaction with peers and mentors all contribute to the willingness to live with long periods of searching for solutions. There is also the young mathematician's extraordinary ability to concentrate, ignore diversions, and follow a sense of direction even in the face of failure. Young women face particularly serious challenges when entering a profession that can isolate them from their peers and family and in which gender bias has been prevalent through most of mathematical history. What makes such risk taking and concentration possible may be the powerful appeal of adding deep, lasting knowledge, knowledge that is fulfilling both aesthetically and intellectually.

Bibliography

Albers, Donald J., & Alexanderson, G. L. (1985). *Mathematical people: Profiles and interviews*. Boston: Birkhauser.

Albers, Don (2007). John Todd—Numerical mathematics pioneer. *College Mathematics Journal 38*(1), 5.

Bell, E. T. (1937). *Men of mathematics*. New York: Simon and Schuster.

Bloom, Benjamin S. (1985). *Developing talent in young people*. New York: Ballantine Books.

Brockman, John (Ed.) (2004). *Curious minds: How a child becomes a scientist*. New York: Pantheon Books.

Chang, Kenneth (2007). Journeys to the distant fields of prime. *New York Times*, March 13, 2007. Retrieved April 10, 2008 from http://www.nytimes.com/2007/03/13/science/13prof.html?_r=1& sq=The%20Mozart%20of%20MAth&st=nyt&oref=slogin&scp =1&pagewanted=print

Csikszentmihalyi, Mihaly, Rathunde, Kevin, & Whalen, Samuel (1996). *Talented teenagers: The roots of success and failure*. New York: Cambridge University Press.

Dyson, Freeman J. (2004). Member of the club. In John Brockman (Ed.). *Curious minds: How a child becomes a scientist*. New York: Pantheon Books.

Feldman, David H. (1986). *Nature's gambit*. New York: Basic Books.

Gallian, Joseph A. (2004). A conversation with Melanie Wood. *Math Horizons 12* (September), 13, 14, 31.

Gustin, William C. (1985) The development of exceptional research mathematicians. In Benjamin S. Bloom (Ed.). *Developing talent in young people*. New York: Ballantine Books, pp. 270–331.

Heims, Steve J. (1982). *John Von Neumann and Norbert Wiener: From Mathematics to the technologies of life and death*. Cambridge, Mass.: MIT Press.

Hersh, Reuben, & John-Steiner, Vera (1993). A visit to Hungarian mathematics. *Mathematical Intelligencer* 15(2) 13–26.

Howe, Michael J. A. (1990). *The origins of exceptional abilities*. Cambridge, Mass.: Basil Blackwell.

James, I. (2002). *Remarkable mathematicians*. Cambridge: Cambridge University Press.

John-Steiner, Vera (1997). *Notebooks of the mind: Explorations of thinking*. New York: Oxford University Press.

Krutetskii, V. A. (1976). *The psychology of mathematical abilities in schoolchildren*. Chicago: University of Chicago Press.

Levinson, N. (1966). Wiener's life. *Bulletin of the American Mathematical Society* 72(1, II), 3, 24, 25.

MacTutor web site. Birkhoff.

Marx, George (1999). The Hungarian gymnasium. *Europhysics News* (Nov./Dec.) p. 130.

Massera, J. L. (1998). *Recuerdos de mi vida academica y política* (Memories of my academic and political life). Lecture delivered at the National Anthropology Museum of Mexico City, March 6, 1998, and published in *Jose Luis Massera: The scientist and the man*. Translated by Frank Wimberly. Montevideo, Uruguay: Faculty of Engineering.

Mordell, L. J. (1971). Reminiscences of an octogenarian mathematician, *American Math, Monthly* 78 952–961,

Morrow, Charlene, & Perl, Teri (Eds.) (1998). *Notable women in mathematics: A biographical dictionary*. Westport, Conn.: Greenwood Press.

Murray, Margaret (2000). *Women becoming mathematicians*. Cambridge, Mass.: Massachusetts Institute of Technology.

O'Connor, J.J., & Robertson, E.F. (2003). George Dantzig http:// www- history.mcs/st-andrews.ac.uk/Mathematicians/Dantzig-George.htm (p.2) (article from MacTutor, School of Mathematics and Statistics, University of St. Andrews, Scotland, JOC/EFR, April 2003.

Olson, Steve (2004). *Count down.* Boston: Houghton Mifflin.

Paulson, Amanda (2004). Children of immigrants shine in math, science. *Santa Fe New Mexican* 813 1/04, p. A5.

Perl, Teri (1978). *Biographies of women mathematicians and related activities.* Menlo Park, Calif.: Addison-Wesley.

Radford, John (1990). *Child prodigies and exceptional achievers.* New York: Harvester Wheatsheaf.

Rathunde, Kevin, & Csikszentmihalyi, Mihaly (1993). Undivided interest and the growth of talent: A longitudinal study of adolescents. *Journal of Youth and Adolescence* 22(4), 385–405.

Reid, C. (1976). *Courant in Göttingen and New York.* New York, Springer-Verlag.

Reid, Constance (1996). *Julia, a life in mathematics.* Washington, D.C.: Mathematical Association of America.

Sacks, Oliver (2001). *Uncle Tungsten: Memories of a chemical boyhood.* New York: Alfred Knopf.

Singh, Simon (1998). *Fermat's Last Theorem.* London: Fourth Estate.

Szanton, Andrew (1992). *The recollections of Eugene P. Wigner as told to Andrew Szanton.* New York: Plenum Press.

Tikhomirov, V. M. (1993). A. N. Kolmogorov. In Smilka Zdravkovska & Peter L. Duren (Eds.). *Golden years of Moscow mathematics.* Providence, R.I.: American Mathematical Society.

Tikhomirov, V. M. (2000). Moscow mathematics 1950–1975. In Jean-Paul Pisier (Ed.). *Development of mathematics 1950–2000.* Boston: Birkhäuser, pp. 1109–1110.

Ulam, S. (1976). *Adventures of a mathematician.* New York: Scribner.

Wiener, Norbert (1953). *Ex-prodigy: My childhood and youth.* New York: Simon and Schuster.

Wigner, Eugene (1992). *The recollections of Eugene P. Wigner as told to Andrew Szanton.* New York: Plenum Press.

Winner, Ellen (1996). *Gifted children: Myths and realities*. New York: Basic Books.

Wolpert, Stuart (2006). Terence Tao, "Mozart of math," is UCLA's first mathematician awarded the Fields Medal, often called the "Nobel Prize in Mathematics." *UCLA News*, August 22, 2006. Retrieved April 10, 2008 from http://newsroom.ucla.edu/portal/ucla/Terence-Tao-Mozart-of-Math-7252

Yau, S. T. (Ed.). (1998). *S. S. Chern: A great geometer of the twentieth century*. Singapore: International Press.

✦2✦

Mathematical Culture

Mathematicians constitute a community with a long, rich history. They recognize each other as fellow members of that community. *What is the culture of mathematics and of the mathematical community?*

This question seems to have been seldom asked and little studied. The American topologist Raymond Wilder was one outstanding contributor to this subject. In this chapter, we attempt to continue his work and highlight some salient features of mathematical culture.

When Paul Halmos was asked, "What is mathematics?" He answered, "It is security. Certainty. Truth. Beauty. Insight. Structure. Architecture."[1] Most mathematicians agree with Halmos and treat the aesthetic components of discovery and proof as a source of joy. Others speak of the study of shapes and patterns. The French mathematician Laurent Schwartz wrote: "I simply found mathematics beautiful—extraordinarily beautiful—and geometry in particular incomparably elegant."[2] Rozsa Péter, the Hungarian author of *Playing with Infinity,* argued that one could do nothing better and more beautiful than to work on mathematics.[3]

In discovering the field in their youth, many mathematicians are delighted by the internal logic of the arguments that lead from axioms to proofs. For many, a general interest in science provides a first entry into rigorous thinking, which later leads them to mathematics after they discover that they lack the dexterity needed for laboratory sciences. (Ralph Boas wrote of his "record bill in broken glassware" when taking chemistry.[4]) Others report that their desire for certainty was not met by physics or biology. One's personal intellectual preferences—or what psychologists call learning styles—must be consonant with the prevalent modes of thought, of the *culture,* and of one's chosen field.

We start with a definition of "culture" from the *American Heritage Dictionary* (2006):

> The totality of socially transmitted behavior patterns, arts, beliefs, institutions, and all other products of human work and thought.
>
> These patterns, traits, and products are considered as the expression of a particular period, class, community, or population.

Culture also encompasses structured domains of human activity with characteristic symbol systems, such as musical or mathematical concepts and notations.[5]

In this chapter, we focus on four aspects of culture. The first is cognitive-emotional. We ask: What habits of mind, behavior patterns and motivation permit young mathematicians to become members of their mathematical community?

The second aspect is aesthetic. Everyone agrees that mathematics can, should, or even *must,* be beautiful. But it is not so easy to explain just what you mean by beautiful mathematics. What patterns, proofs, and discoveries do mathematicians call beautiful?

The third aspect is social. What shared values, stories, traditions, and institutions support the discipline? Who are the insiders and who are the outsiders?

The fourth aspect of mathematical culture we look at is how it deals with internal tensions: How does the discipline identify and address conflicts? How is a discoverer rewarded? How are prizes shared? How does competition coexist with collaboration?

Mathematical Cognition and Feelings

The acquisition of abstract modes of thought is basic to mathematical activities. It starts early in children's lives. For instance, at birthday parties, preschoolers hold up four fingers, thereby showing that they understand the equivalence relation of whole numbers. They acquire more abstract notions when they learn number words and symbols.

Davis and Hersh describe two aspects of abstraction. One is "idealization," that is, stripping away irrelevant details, such as the thickness of the pencil line with which a triangle is drawn. The second is "extraction," that is, pulling out the essential features of a problem, for instance, representing a complicated maze as a graph by letting selected positions in the maze become nodes on the graph.

Mathematics progresses, both historically and psychologically, by successive expansions: counting is followed by geometric and arithmetic operations, and then by the study of more general patterns. The connection between a pattern in nature and a mathematical representation has powerful emotional appeal. It is part of the emotional and aesthetic appeal of mathematics. This connection was illustrated by Steven Strogatz's experience with the parabola described in the previous chapter.

Participants in mathematical culture create symbols and notations, which constitute the shared tools or "language" of their community. By the use of symbols and notations, concepts are clarified and made precise. After a symbol is invented by an individual, it is appropriated, evaluated, disseminated, and applied by many mathematicians. The dynamic connection between individual cognition and communal activity takes place through conversations, academic apprenticeships, publications, conferences, and textbooks.

In ancient times, new symbols were created to meet the needs of commerce and astronomy. Scribes used Greek letters for numbers to replace the earlier reliance on tallies and clay tokens. Today, new concepts and symbols still arise from the challenges of the physical sciences as well as from the evolving internal logic of mathematics.

The evolution of the symbolic tools of mathematics has been uneven. Long periods of stasis (for instance, during the Middle Ages in Europe) were followed by periods of rapid progress. Societies with active commerce, broad geographic contacts, and emerging technologies provide a generative environment for mathematical growth. These contrast sharply with the stability of static isolated communities. An example of an extremely isolated community is the Pirahã, a small tribe of hunters and gatherers who live in the Amazon basin. The linguist Daniel Everett spent years studying the Pirahã. He found that they have no arithmetic and that their language has no number words. They communicate almost as much by singing, whistling, and humming as by using consonants and vowels. Everett suggests that their lack of number words and counting, and also a lack of cultural stories, are associated with a certain pervasive value in their conversations. The Pirahã avoid talking about any knowledge that goes beyond personal, immediate experience.[6]

The lack of number words is not limited to the Pirahã. There are hundreds of indigenous languages in Papua, New Guinea, many of which have words only for "one," "two," and "many." Cross-cultural and historical analyses suggest that what we consider universal in our experience with numbers may actually be specific to societies that are economically developing and culturally open.

The habits of mind *within* a culture are also diverse. Even within the contemporary Western mathematical community, logic and abstraction are not the only ways that mathematicians think. They are supplemented by intuition, analogy, and visualization.

Benoit Mandelbrot is famous for his work on fractals. When he was a young child, his uncle taught him to play chess, to study maps, and to read very fast.[7] As a student he would rely on shapes to represent mathematical functions. When he reached university age, he managed to pass the very challenging entrance examinations to France's elite higher educational institutions, the École Normale Superieure and the École Polytechnique. He writes, "Faced with some complicated integral, I instantly related it to a familiar shape; it was usually the exact shape that motivated this integral. I knew an army of shapes I'd encountered once in some book or in some problem, and remembered forever, with their properties and peculiarities."[8]

Mandelbrot's strong use of visualization went against the prevailing academic trends of the postwar decades. The visualization described by Mandelbrot is one form of mathematical intuition, by which we mean a perception that is "plausible or convincing in the absence of proof."[9] Such intuitive perceptions are essential in discovery; they contrast with the more rigorous, deductive methods needed for justification.

For some students who find mathematics frustrating in school, the monotonous repetition of drill in algorithms contributes to a sense of alienation. In contrast, when a teacher takes a broader approach, including problem solving, guessing, and intuition, learners can engage in a freer, gamelike experimentation with ideas. They think and communicate inferentially through examples and visualization, as well as by logical rules.

Thought is hard to observe. One way to learn about it is by introspection. In Jacques Hadamard's famous essay, "The Psychology of Invention in the Mathematical Field," he asked his colleagues, "What internal or mental images, what kind of 'internal words,' do mathematicians make use of; whether they are motor, auditory, visual, or mixed, depending on the subject which they are studying?"[10] Einstein's answer has been frequently quoted. He wrote: "The words or the language, as they are written or spoken, do not seem to play any role in my mechanism of thought. The psychical entities which seem to serve as elements in thought are certain signs and more or less clear images which can be 'voluntarily' reproduced and combined."[11]

Hadamard's own mode of thought was similar to Einstein's. He relied on mental pictures to make discoveries and had a

Figure 2-1. André Weil. Source: *More Mathematical People: Contemporary Conversations*. Eds. Donald J. Albers, Gerald L. Alexanderson, and Constance Reid.

challenging time translating them into words. This tendency made him particularly interested in visual, intuitive, and unconscious approaches to discovery in mathematics, which would be followed by the verifying and "précising" stage.

One description of rapid thought processes that emerge from an unknown source, what some call the "Aha" experience, is provided by André Weil, one of the founders of Bourbaki. (This group of French mathematicians reformulated the basis of their discipline in the 1930s and wrote collaboratively under their chosen name.) He writes, "Every mathematician worthy of the name experienced, if only rarely, the state of lucid exaltation in which one thought succeeds another as if miraculously, and in which the unconscious (however one interprets this word) seems to play a role. . . . Once you have experienced it, you are eager to repeat it but are unable to do so at will, unless perhaps by dogged work which it seems to reward with its appearance."[12]

The creativity researcher Mihaly Csikszentmihalyi describes the state of deep immersion as "flow," a condition when challenge and skill are well matched. In this state, people concentrate so deeply on a task that "they stop being aware of themselves as separate from the actions they are performing."[13]

Mentors and advisers model both the cognitive and motivational aspects of mathematical life. As models, they teach persistence. Researchers persist in lengthy and at times frustrating labor, which may be followed by relief and joy. In 1927 Oscar Zariski had recently emigrated from Italy to America. He wrote letters to his wife in which he shared the roller coaster of emotions that frequently accompany intense mathematical work: "Now I am going through a feverish period. . . . If I will succeed in bringing this research to a good outcome . . . it will be an affirmation of my worth to the professors who have been vitally interested in this problem."[14] Weeks later, he was still nailed to his desk, still unable to solve the problem: "Ah, well, I have had a very bad period. I have gotten stuck on a hard point that I have not been able to overcome for weeks. . . . I have been close to despair, even though Prof. Coble consoled me, saying that one must always take these issues philosophically. And that mathematics is 'a slow game.'"[15] It took him another month to complete the research by using topological methods in algebraic geometry. This work led him to meet the famous topologist Solomon Lefschetz, with whom he shared his Russian Jewish heritage.

Creativity requires continuity of concern and intense awareness of one's active inner life. Routine activities can be useful as a backdrop to the researcher's deep immersion in his or her problem. The British mathematician (of Lebanese and Scottish origins) Michael Atiyah described some of his problem-solving ways:

Figure 2-2. Michael Atiyah (center), English geometer and mathematical physicist, with friends. Courtesy of Dirk Ferus.

I think that if you are actively working in mathematical research, then the mathematics is always with you. When you are thinking about problems, they are always there. When I get up in the morning and shave, I am thinking about mathematics. When I have my breakfast, I am still thinking about my problems. When I am driving my car, I am still thinking about my problems. All in various degrees of concentration. . . . There are occasions when you sit down in the morning and start to concentrate very hard on something. That kind of acute concentration is very difficult for a long period of time and not always very successful. Sometimes you will get past your problem with careful thought. But the really interesting ideas occur at times when you have a flash of inspiration. Those are more haphazard by their nature; they may occur just in casual conversation. You will be talking with somebody and he mentions something and you think, "Good God, yes, that is just what I need . . . it explains what I was thinking about last week." And you put two things together, you fuse them and something comes out of it.[16]

At its most rewarding, mathematics provides aesthetic enjoyment, the satisfaction of clear and deep thinking, and the emotional roller coaster of discovery.

A prevalent myth about mathematical thinking is that somehow it is all the same kind of thing. "Either you are mathematically inclined or you're not." Bur there are more than one kind of mathematical talent and mathematical thinking. For those who want straightforward predictions of academic or professional success, tests like the IQ test are appealing. But the development and application of talent is quite complex. Some individuals reveal interests and skills at an early age. As a child, John von Neumann was intrigued by his grandfather's ability to rapidly perform complex mathematical calculations (a skill for which von Neumann later became famous). He loved arithmetic, was good in classical and modern languages, and had an extraordinary memory.[17]

In contrast, Roger Penrose, today a preeminent leader in geometry and relativity theory, as a child loved geometry but was slow in arithmetic. In fact, he was moved back a class in grade school. "You had to add up, add and multiply, and do various things, and I just got lost—I always got lost in the middle. I was very slow, I could do these things, but only at great length, thinking about them. I suppose people tended to do these things automatically, whereas I had to figure it out each time. It was probably a help in the long run, but at the time it was a disadvantage."[18] These two mathematicians present a contrast akin to that between Mozart and Beethoven. The prodigious ease, fluency, and productivity of the young Mozart are akin to that of von Neumann. Penrose is more like the hard-working Beethoven, who constructed his music laboriously over a long period of time.

In chapter 1 we describe the varied ways that future mathematicians first discover their fascination with numbers and patterns. Some, like von Neumann, are highly focused prodigies. Another example is the recent Fields Medal winner Terence Tao, who enjoyed numbers at a very young age. He learned math and science very fast, but he did not like to write essays. "I never really got the hang of that." And he said: "These very vague, undefined questions. I always liked situations where there are very clear rules of what to do." Assigned to write a story about what was going on at home, Terry went from room to room and made detailed lists of the contents.[19]

But other future mathematicians have had broad interests, including music and poetry. They decided on a career in a more deliberate way, making a carefully considered choice as young adults.

Diversity in style is also present in mature mathematicians. Davis and Hersh describe how they differ in their reliance on "analytical" versus "analog" processes. In the former, symbolic and verbal processes play a leading role, while in the latter, geometric and visual approaches are more important. Davis recalls spending considerable time on a problem on the "theory of functions of a complex variable."[20] The problem could be studied from a geometric or from an analytical point of view or by combining the two. Davis found that as he worked on the problem, textbook illustrations of spheres, maps, and surfaces came to mind, along with a certain melody:

> I worked out, more or less, a body of material which I set down in an abbreviated form. Something then came up in my calendar which prevented me from pursuing this material for several years. I hardly looked at it in the interval. At the end of this period, time again became available, and I decided to go back to the material and see whether I could work it into a book. At the beginning I was completely cold. It required several weeks of work and review to warm up the material. After that time I found to my surprise that what appeared to be the original mathematical imagery and melody returned, and I pursued the task to a successful completion.[21]

This quote not only shows the diverse modalities of thought that accompany mathematical problem solving, it also reveals

Figure 2-3. Irving Kaplansky, Don Albers and Shiing-Shen Chern. Source: *More Mathematical People: Contemporary Conversations*. Eds. Donald J. Albers, Gerald L. Alexanderson, and Constance Reid.

a form of cognitive shorthand. Davis's observation of his own thinking process adds an interesting dimension to our understanding of inner thought processes.[22]

Thus, mathematicians are not limited to a single mode of thought. They rely on intuition, logic, visual and verbal processes, inferences, and guesses. They are assisted by the many artifacts and symbol systems of their profession.

Within pure mathematics, a distinction has been made between theorizers and problem solvers. The great names from the history of mathematics—Newton, Euler, Gauss, Riemann, Cantor—are revered to this day for creating theories that are still the mainstay and generating source of mathematics. On the other hand, a mathematician is most likely to become instantly famous in his own time if he or she solves a famous problem. The problem should have been outstanding for a long time, and it should have withstood attack from many famous mathematicians. But the greatest mathematicians must be counted among both groups, they were both theorizers and problem solvers.

Italian-born mathematician and philosopher Gian-Carlo Rota
(1932–1999) has written incisively and wittily:

> To the problem solver, the supreme achievement in mathematics
> is the solution to a problem that had been given up as hopeless.
> It matters little that the solution may be clumsy; all that counts
> is that it should be the first and that the proof be correct. Once
> the problem solver finds the solution, he will permanently lose
> interest in it, and will listen to new and simplified proofs with an
> air of condescension suffused with boredom.
>
> The problem solver is a conservative at heart. For him math-
> ematics consists of a sequence of challenges to be met, an ob-
> stacle course of problems. The mathematical concepts required
> to state mathematical problems are tacitly assumed to be eternal
> and immutable.
>
> . . . The problem solver is the role model for budding young
> mathematicians. When we describe to the public the conquests
> of mathematics, our shining heroes are the problem solvers.
>
> To the theorizer, the supreme achievement of mathematics is
> a theory that sheds sudden light on some incomprehensible phe-
> nomenon. Success in mathematics does not lie in solving prob-
> lems but in their trivialization. The moment of glory comes with
> the discovery of a new theory that does not solve any of the old
> problems but renders them irrelevant.
>
> The theorizer is a revolutionary at heart. Mathematical con-
> cepts received from the past are regarded as imperfect instances
> of more general ones yet to be discovered. . . . Theorizers often
> have trouble being recognized by the community of mathemati-
> cians. Their consolation is the certainty, which may or may not
> be borne out by history, that their theories will survive long after
> the problems of the day have been forgotten."[23]

So we can describe mathematicians as theorizers versus problem solvers or as those who prefer algebraic methods to geometric methods. Some are more rigorous, some more intuitive.
But none of these categories is fixed. There are researchers in

between who sometimes work one way or shift to another. By considering these varied styles, we are led to recognize and appreciate the great diversity of thinking modes that contribute to the experience of mathematicians.

In challenging the myth of homogeneity, we open the field to groups and individuals with varied talents and practices. Such recognition has both an emotional and an educational benefit. It supports the varied cognitive styles of potential members of the profession. It also validates innovative practices in reform classrooms, where young learners are encouraged to experiment with various problem-solving strategies.

Mathematical Beauty

Most mathematicians agree that the aesthetic components of discovery and proof are a source of joy for researchers and practitioners in the field. Addressing the British Association for the Advancement of Science in 1885, Arthur Cayley said, "It is difficult to give an idea of the vast extent of modern mathematics. I mean extent crowded with beautiful detail—not an extent of mere uniformity such as an objectless plain, but of a tract of beautiful country seen at first in the distance, but which will bear to be rambled through and studied in every detail of hillside and valley, stream rock, wood, and flower."[24] David Ruelle (the French contributor to chaos theory) writes, "The beauty of mathematics lies in uncovering the hidden simplicity and complexity that coexist in the rigid logical framework that the subject imposes."[25] He has given a strikingly perceptive explanation: "The infinite labyrinth of mathematics has thus the dual character of human construction and logical necessity. And this endows the labyrinth with a strange beauty. It reflects the internal structure of mathematics and is, in fact, the only thing we know about this internal structure. But only through a long search of the labyrinth do we come to appreciate its beauty; only through long study do we come to taste fully the subtle and powerful aesthetic appeal of mathematical theories."[26]

In a famous essay, the British mathematician G. H. Hardy (1877–1947) gave three criteria for beautiful mathematics. He said it must be serious, deep, and surprising. *Serious*, he said, in contrast with chess problems, for example, which are a kind of mathematics but not serious mathematics. *Deep*—for example, he said the existence of infinitely many primes, as proved by Euclid, is not deep; in contrast, the prime number theorem, which gives the proportion of primes among all whole numbers, is deep. And finally—*surprising*, which even more than deep or serious is clearly subjective. Whether the reader is surprised depends on what the reader already knows.

To explain to the lay person what he meant by beauty in mathematics, Hardy gave two examples from Euclid that he thought were exemplary: "No fraction when squared can be equal to 2" and "Given any finite collection of prime numbers, it is always possible to find a larger prime not included in the collection." Or, in brief, "The square root of 2 is irrational" and "There are infinitely many primes."[27]

Gian-Carlo Rota also attempted to explicate mathematical beauty. Unlike Hardy, he didn't think "surprising" was necessary for "beautiful." After listing many specific examples, he concluded that beauty is "enlightenment." We agree that a sense of enlightenment can make one exclaim, "Beautiful!" But Rota didn't make much of an attempt to explicate what "enlightenment" is. Surprise, depth, simplicity, clarity, and directness all contribute to the aesthetic experience of mathematics.

An early member of Harvard's mathematics faculty, George David Birkhoff (1884–1944) published a mathematical theory of aesthetics. Although few people take it seriously as a contribution to philosophy, Birkhoff did attempt to think carefully about what is the basis of aesthetic pleasure. He proposed a formula $E = M/C$. Here E is aesthetic pleasure and is increased by maximizing M, the richness or inclusiveness of the artistic object, and minimizing C, the complexity of means. Although this algebraic formula may not be of much use, the two factors M and $1/C$ in it have been identified by other authors, as "integration" versus "simplicity." Another way to think about this relationship is to recognize that integration is a form of simplicity.

It is useful to connect aesthetics in mathematics to aesthetics in other artistic realms. Of course, the geometric or visual parts of mathematics have well-known connections to visual art, first of all to perspective drawing and painting, and then as sources of images for artists to use in any way they wish. Drama and theater provide useful viewpoints from which to judge mathematical presentations. Certainly an effective speaker at a colloquium or seminar can use theatrical or dramatic effects and devices to enhance his or her presentation. The analogy between pure mathematics and music is even more natural. They both communicate patterns that speak for themselves, requiring no verbal justification or amplification.

Could the criteria for judging a story or a musical composition be relevant for judging a mathematical paper? Coherence; unity; arousal of interest at the beginning; a satisfying sense of completion at the end; use of recurrent themes or leitmotifs to connect the beginning, middle, and end; logical or comprehensible connection from one chapter or one movement to the next—the presence or absence of these qualities can help one to make an aesthetic evaluation. The same qualities can be sought in a mathematical talk, article, or book. This would lead to an aesthetic evaluation of the presentation of some mathematics. On the other hand, it would be a different matter to evaluate the mathematical *content*. And it might not be easy to separate presentation from content. Given two different proofs of the same result, one mathematician might call them mathematically equivalent, differing only in presentation; another might say that the differences between them are mathematically significant.

The question remains, how to aesthetically judge mathematical content itself. We think we can point to three important elements of beauty in mathematical content: simplicity, concrete specificity, and unexpected or surprising integration or connection of disparate elements. The Bourbaki style was ultimately unsatisfying because it bought simplicity at the expense of concreteness and multiconnectedness.

In a much quoted sentence, Hardy wrote, "The mathematician's patterns, like the painters or the poet's, must be *beautiful*; the ideas, like the colours or the words, must fit together in a harmonious way. Beauty is the first test; there is no permanent place in the world for ugly mathematics."[28] This is easily refuted. Hardy was wrong! It is not hard to find permanent parts of mathematics that no one thinks are beautiful. Rota's article contains examples. At an elementary level, we may mention the "quadratic formula,"

$$x = [- b \ (+ \text{ or } -) \ \sqrt{b^2 - 4ac}]/2a,$$

which solves the general quadratic equation

$$ax^2 + bx + c = 0.$$

This is one of the most memorized formulas in math. Not beautiful!

The truth is that a mathematician striving to solve a problem does not worry much about making it come out beautiful. The aesthetic judgment comes in at the beginning, when she chooses a problem to solve or an area to work in. Yes, at that stage, beauty attracts and ugliness repels. But once the battle is joined, anything that works is welcome. Later, after the proof is found and the result is known, one looks back and tries to refine and beautify the proof. Maybe other mathematicians also will pick up the problem and look for a different, more attractive approach. But there is no assurance that an important mathematical result will yield to a beautiful attack.

A notorious example of unbeautiful proof is computer proof. Starting with the four-color theorem, and since then followed by quite a few other examples, mathematicians have found that what they cannot do by hand they can sometimes do by machine. In 1976, when Kenneth Appel, Wolfgang Haken, and

John Koch published their computer proof of the four-color theorem, Tom Tymoczko and others thought that this raised a question about the nature of mathematical proof. But it seems today that mathematicians generally accept a well-verified, stable, robust computer proof as a real proof. That is to say, they accept the truth of a theorem like the four-color theorem on the basis of the computer's testimony. (The four-color theorem says that four colors are sufficient to color any map with no two adjacent countries having the same color.) But there is still a serious objection: "Yes, you proved it by computer, but I don't like it!" Computer proofs seem to be an unwelcome last resort, an ugly thing that does the job. What we really want is something beautiful, not something ugly. On the other hand, some mathematicians find the work of discovering and improving computer proofs to be intellectually attractive. They can look at a neat piece of computer code and think, "That's beautiful!"

Johann Lambert's near-discovery of non-Euclidean geometry is a remarkable example of the aesthetic sense in conflict with the seeming demands of logic. Lambert (1728–1777) was a forerunner of Gauss, Nikolai Lobachevsky (1792–1856), and János Bolyai (1802–1860), struggling with the "problem of parallels" that gave birth to non-Euclidean geometry. He was a younger contemporary of Euler and became the leading German mathematician of his time. Although he ended up mistakenly rejecting non-Euclidean geometry as contradictory, he was very strongly tempted by the beautiful properties such a geometry would have. "There is something exquisite about this consequence, something that makes one wish that the third hypothesis were true! But all these are arguments dictated by love and hate, which must have no place, either in geometry or in science as a whole."

This confession is a fascinating manifestation of the emotional side of mathematical life. We disagree with Lambert's view that "love and hate have no place in geometry." The presence of love and hate in mathematics is unavoidable—more love than hate. Such emotions are especially evident in mathematical discovery and mathematical aesthetics. Yet, of course, Lambert was right to accept consistency as the final arbiter of a mathematical concept.

Social Aspects of Mathematical Culture

From the perspective of an outsider, mathematicians appear to be solitary thinkers. The amount of collaboration that takes places between them is rarely understood or publicized. But at closer acquaintance these abstract thinkers share a lot in their work and also in their common mathematical histories.

Young mathematicians refine their abilities to work on problems and theorems through apprenticeships. An essential part of graduate education includes closely watching a more mature mathematician engaged in mathematical practice. The Hungarian-born Paul Halmos (1916–2006) wrote of his advisor Joseph Doob, with whom he studied at the University of Illinois: "He has the one absolutely necessary quality of a teacher: he can see what it is that the learner is not seeing."[29]

Conversations with one's peers can also greatly contribute to mathematical habits of the mind. Halmos describes his rapport with his friend Allen Shields:

> The best seminar I ever belonged to consisted of Allen Shields and me. We met one afternoon a week for about two hours. We did not prepare for the meetings and we certainly did not lecture at each other. We were interested in similar things, we got along well, and each of us liked to explain his thoughts and found the other a sympathetic and intelligent listener. We would exchange the elementary puzzles we heard during the week, the crazy questions we were asked in class, the half-baked problems that popped into our heads, the vague ideas for solving last week's problems that occurred to us, the illuminating problems we heard at other seminars—we would shout excitedly, or stare together at the blackboard in bewildered silence—and, whatever

we did we both learned a lot from each other during the year the seminar lasted, and we both enjoyed it.[30]

One of the most carefully documented mathematical discoveries is the work of Andrew Wiles. To prove Fermat's Last Theorem, he devoted seven long years working in the attic of his house. He relied upon the work of many of his predecessors but sustained his own momentum. As he approached the culmination, he felt the need for communication. He recruited his colleague Nick Katz as an interlocutor. In June of 1993 Wiles presented his proof at a conference in Cambridge, England (his alma mater). The announcement of the proof galvanized the international mathematical community. During the extensive review process, a gap was found that forced Wiles to go back to work. Wiles decided to go into complete seclusion. "As months went by, Dr. Wiles finally decided that he needed help. He was frightened that he would mislead himself. Spending years, or decades going in circles trying frantically to fix the gap with methods that never would work. . . . 'You get caught in problems, trapped in them. I knew the dangers psychologically . . . I needed someone to talk to all the time . . . I wanted someone I was sure of.' He chose a former student of his, Richard Taylor of Cambridge University."[31] Fortunately, Taylor was ready for a sabbatical leave and was able to join Wiles in Princeton. It is interesting that at a crucial stage a collaborator was sought, even by a man committed to solving a problem by himself, a problem that had been of interest to him throughout his entire life, starting at the age of 10. Then, with the added perspective and energy of a scientific partner, he successfully resolved one of mathematics oldest challenges.

During Wiles' 8-year ordeal he brought together virtually all the breakthroughs in 20th century number theory and incorporated them all in one mighty proof. He created completely new mathematical techniques and combined them with traditional ones in ways that had never been considered possible. In doing so he opened up new lines of attack on a whole host of other problems.[32]

Figure 2-4. Andrew Wiles, who proved Fermat's last theorem. Courtesy of C. J. Mazzochi.

The Love of Stories

Another social feature of mathematics is telling the personal stories of accomplishments and eccentricities of mathematicians, which they discuss with great gusto when colleagues meet. Often these stories tell of sustained immersion in intense research mathematics.

Isaac Newton (1643–1727) described how he made his discoveries: "I keep the subject constantly before me and wait until the first dawnings open little by little into the full light."[33] It was hard for Newton to switch his attention away from mathematics, even when he was hosting a party. His friend William Stukeley recalled that when he had friends in his rooms, if he went into his study for a bottle of wine and a thought came into his head, he would sit down to write and forget his friends.[34]

Norbert Wiener (1894–1964) of MIT was well known as an extreme example of someone who could get lost in thought. Once while walking on campus, Wiener met an acquaintance, and after a while he asked his companion: "Which way was I walking when we met?" The man pointed, and Wiener said, "Good. Then I've had my lunch."[35]

Another example of absent-mindedness is John Horton Conway of Princeton University, inventor of surreal numbers and the game of "Life." Conway's office is incredibly chaotic:

There were papers and books scattered everywhere. Puzzles, games, charts, novelties, and other paraphernalia were piled and heaped and stacked on all horizontal surfaces. Conway realized he had a problem when he could no longer lay his hands on his latest theorem, or list of problems or new conjectures. He set about to design a physical device that would address his quandary, and imbue some order into the chaos. After some time and effort, he had produced a set of plans. He was about to go off and find a craftsman to implement his idea in wood and metal when he noticed that such an item was already standing, empty and unused, in the corner of the office. It was the filing cabinet![36]

It is hard to choose from the hundreds of stories illustrating the depths of engagement of mathematicians. Here is one of our favorites. It is about the topologist R H Bing (1914–1986), a famous student of Robert Lee Moore (1882–1974). (The R and H are not initials, Bing's complete name consisted of the two letters R and H and his family name Bing.)

One day R H Bing was driving a group of mathematicians to a conference. As usual, Bing launched into a detailed discussion of

Figure 2-5. John Conway, British-American algebraist and number theorist, with a friend in 1987. Courtesy of Archives of the Mathematisches Forschungsinstitut Oberwolfach.

a particular problem in topology that he wanted to solve. The passengers were nearly as interested in the math as Bing, but they were rather nervous because the weather was bad, visibility poor, and there was lots of traffic. Bing did not seem to be giving the matter his full attention. To make matters worse, the windshield was completely fogged over and it was nearly impossible to see. A sigh of relief was quietly breathed when Bing began to

Figure 2-6. R H Bing and Mary Bing. Courtesy of Dolph Briscoe Center for American History, The University of Texas at Austin. Identifier: di_05557. Title: Bings. Date: 1975/08. Source: Halmos (Paul) Photograph Collection.

reach toward the windshield. Everyone supposed that he was going to wipe the windshield clear and pay attention to what was happening on the dark and stormy road. Instead Bing started drawing diagrams with his finger in the windshield mist.[37]

These narrative exchanges provide the social glue to hold a diverse and dispersed community together. In cherishing their colleagues' foibles, mathematicians come to terms with the emotional complexity of living in an abstract world.

Through most of history, the world of mathematics has been separate from the broader cultural milieu. But recently the lives of mathematicians have become of increasing popular interest. There was *A Beautiful Mind*, a successful book and movie about

John Nash, and then the Broadway hit play called *Proof*; and articles about mathematicians in magazines such as *The New Yorker*. These biographical and fictional representations focus on the intensity of mathematical life and the consequences of such dedication. Other themes of these stories are competition in mathematical discovery and the vagaries of success. The main characters in *Proof* are a retired University of Chicago mathematics professor and his daughter. In the last years of his life, the professor kept notebooks of his mathematical explorations, and no one knew what was in them. The struggle for possession of these notebooks, and the actual identity of their author, are a substantial part of the play, delving into issues of ownership, gender, and fame.

Tensions and Their Resolution

In general, mathematicians are supportive of each other as they focus on solving problems. The rules serve to decide when a problem has been solved, and there are not as many battles for credit in mathematics as in archeology or linguistics. The celebration of Wiles' success in proving Fermat's Last Theorem was widespread and joyous; it is an instance of the usual mutual support among mathematicians.

Yet there are exceptions to this camaraderie. Prizes, academic politics, and the value of fame and visibility in the broader culture inevitably impact the discipline. We will talk about three dramatic examples. First, there is Grigori Perelman's sensational proof of the Poincaré conjecture (which says that under a certain simple condition, any three-dimensional manifold is continuously deformable to a sphere) and the involvement therein of the great Chinese geometer Shing-Tung Yau and two of his students. Then there is the tangled tale of the recognition of chaos as a major, nearly ubiquitous feature of nature. And finally, there is the controversy in the mathematics department at the University of California at Berkeley about the denial and then the granting of tenure to the American mathematician Jenny Harrison.

Poincaré first stated his conjecture in 1904. The conjecture was proved in dimensions 5 and greater by Steven Smale and in dimension 4 by Michael Freedman. Surprisingly, a solution in three dimensions proved to be much more difficult than in the higher dimensions.

As with many important mathematical problems, progress was cumulative. William Thurston, now at Cornell, formulated what is known as the "geometrization conjecture," which implies the three-dimensional Poincaré conjecture. Richard Hamilton of Columbia University made a major contribution by introducing the method of the "Ricci flow" to the geometry of manifolds. In the early 1990s, after making a presentation at Berkeley, Hamilton was approached by a shy Russian, Grigory Perelman, who was spending a couple of years in the United States. At their first meeting, Perelman was just asking questions. Later he made some contributions to Hamilton's work, the importance of which was not immediately obvious to Hamilton. While Perelman was still in the United States, he made some additional progress on a particularly difficult aspect of the problem. As his reputation grew, he received several job offers in the United States. Although he would have enjoyed collaborating with Hamilton, his overtures were ignored. He decided to return to St. Petersburg. Using the Internet, he could work alone while continuing to tap a common pool of knowledge.[38]

Hamilton was also an important influence on the famous geometer Shing-Tung Yau, whose career accomplishments include a Fields Medal, appointments at Harvard and the Institute for Advanced Study, and an honorary professorship in mainland China. Yau and Perelman are vastly different in their ways of dealing with recognition and public acclaim. "Grisha" Perelman lives very simply with his mother and connects to mathematics via the Internet. While his work is highly respected, both in his home country and abroad, he has become a recluse during the last few years. In contrast, Yau is a very public figure whose

Figure 2-7. Grisha Perelman, who proved Poincaré's conjecture. Courtesy of ICM2006 Madrid.

energy, tenacity, and ambition are widely known. He grew up in Hong Kong, received his Ph.D. at Berkeley, became famous by proving the Calabi conjecture, and during his tenure at Harvard trained dozens of young mathematicians. He has also played a prominent part in China's recent efforts to strengthen scientific research and education.

Perelman sent Yau an e-mail in November 2002 describing some of his results, but he did not receive a response. Subsequently Perelman started to publish aspects of his proof for the Poincaré conjecture on the web site "arXiv.org." He provided procedures that overcame the blocks in Hamilton's approach. Many mathematicians were impressed and hoped that the conjecture had been solved. Perelman made presentations in the United States about his new results. While traveling on the East Coast, he was eager to engage Hamilton in discussing his new work, but he was unsuccessful. In an article in *Science,* Yau pointed out that parts of Perelman's proofs were sketchy. But others, including the team of Gang Tian and John Morgan, checked Perelman's work and found it acceptable. Once the work was verified, Perelman was offered the Fields Medal.

Yau encouraged two of his younger colleagues, Huai-Dong Cau and Xi-Ping Zhu, to write up their own extended presentation of the proof for the *Asian Journal of Mathematics,* of which Yau is the principal editor. The publication was unusually hasty

Figure 2-8. Two geometers: Robin Hartshorne of Berkeley and Shing-Tung Yau of China. Courtesy of Gerd Fischer.

and seemed timed to reflect some degree of glory or credit on China. In a rare gesture, Perelman declined the offer of the Fields Medal. He said, "Everybody understood that if the proof is correct then no other recognition is needed."[39] It is difficult to know how much the controversy over credit contributed to Perelman's decision, and whether his withdrawal from mathematical activity is permanent. In the meantime, Yau has threatened to sue *The New Yorker*. They published an extensive interview with Perelman accompanied by an offensive cartoon showing Yau pulling the Fields Medal from Perelman's neck. Several well-known mathematicians filed letters with Yau's lawyers testifying to Yau's devotion to and accomplishments in mathematics.

This controversy illustrates an interesting distinction between "intrinsic" and "extrinsic" motivation that is made in the creativity literature by Teresa Amabile. "Intrinsic motivation" refers to the enjoyment and satisfaction that a person derives from engaging in the creative activity.[40] For a mathematician, that includes the enjoyment and satisfaction from his or her colleagues' appreciation that the work is correct, interesting, and important. Perelman's comment above is a beautiful example of intrinsic scientific motivation; his reward is simply his success in solving the problem.

On the other hand, there can be motivation for "extrinsic" recognition—promotions, pay raises, prizes, power, and prestige in the wider community. In the case of the Poincaré conjecture, there is actually a million-dollar prize, offered by the Clay Foundation. Are these extrinsic motivations supportive, or do they interfere with creativity? Some researchers argue that such rewards may contribute to task-focused motivation, increasing creative individuals' level of concentration.[41] But Amabile's original work suggested the contrary—that extrinsic rewards may divide one's attention between the creative task itself and the extrinsic goals. More recently, she has proposed the possibility of an interactive effect of the two types of motivation. In Yau's case, this latter model seems applicable. His long, deep involvement with challenging mathematical problems is driven by intrinsic motivation. At the same time, in contrast with Perelman, Yau also thrives on public success and influence and refers with great pride to his various awards.

Which of these two men represents mathematical culture? In the view of the public, where mathematicians are seen as introverted and unconventional, Perelman fits the image. But with the discipline becoming increasingly a part of social life, with broad-ranging discussions of mathematics' usefulness and with policy debates and cinematic depictions, Yau fits a second view of the culture—one of political relevance within academia and the larger society. This story raises the issue of commonality and diversity within the culture. To be a successful mathematician requires focused energy, problem-solving skills, persistence, creativity, and the effective use of logic. Enthusiasm for the beauty and clarity of mathematical problems unites mathematicians, who enjoy discussing great discoveries, and the varied aspects of their joint history. At the same time, differences in temperament and in the desire for public acclaim can lead to fractures in a community that is inherently interdependent. There is a continuum, of degree of concern with extrinsic versus intrinsic scientific rewards, in mathematics; Yau and Perelman occupy two widely separated positions on that continuum.

As mathematics is not an empirical science, agreement on the procedures governing the acceptance of proofs is crucial to the profession. A new finding is normally submitted to a journal. Its editors then assign "referees"—specialists to determine whether it stands up under expert scrutiny. In some cases of unexpected or very complex results, the process of verification may become contentious. This was illustrated in the Perelman-Yau controversy recounted above.

Occasionally, a mathematician may decide to forego publication of a dramatic new discovery. This was the choice made by Carl Friedrich Gauss (1777–1855) when he was working on Euclid's parallel postulate. Gauss discovered a new geometry, completely different from that of Euclid. It was later independently discovered by the Russian Nikolai Lobachevsky and the Hungarian János Bolyai. But Gauss refrained from publishing his findings (of what later became known as non-Euclidean geometry) because he disliked controversy and feared that he would be attacked on philosophical grounds by some of the followers of Immanuel Kant. This new geometry, as developed by Lobachevsky and Bolyai, was ignored for a long time by mathematicians, only to later find its usefulness proven mathematically in Poincaré's function theory, and physically in Einstein's general relativity.

Our second example shows the difficulties that can arise when a new finding violates prevailing mathematical intuition. Against all expectations, it turns out that, in higher than two dimensions, "almost all" dynamical systems are unpredictable in the long run—they are chaotic. The first inkling of this fundamental feature of the physical universe came in Poincaré's work on the stability of the solar system. (This is the most famous example of an n-body problem.)

In honor of the 60th birthday of King Oscar II of Sweden and Norway, *Acta Mathematica* offered a prize for the best essay on this celebrated problem. Poincaré's contribution, including the notion of "characteristic exponents," had a lasting impact on the study of dynamics. His qualitative, topological approach to this problem is a wellspring of topology, which became one of the main branches of mathematics in the 20th century. In Poincaré's analysis of the *n*-body problem he discovered what he called the "homoclinic tangle." In three and higher dimensions of phase space, a trajectory can loop around itself in an infinitely complex entanglement which he found it hopeless to sort out. This was the first mathematical manifestation of the phenomenon now called chaos.[42]

Poincaré recognized the importance of this discovery but did not try to explore it in fuller detail. The full story was told by June Barrow-Green, a math historian, who wrote: ". . . [Poincaré] had found this result so shocking that it is not perhaps surprising that he did not consider the possibility of even more complex solutions."[43] For the next 30 years, this insight of Poincaré was largely ignored, almost forgotten. When Hadamard and Cartan developed some of Poincaré's ideas, they ignored chaos. It contradicted the dominant belief that mathematics and science enable us to control nature.

The 2-body problem—the motion of the earth around the sun, ignoring the perturbing effects of the other planets, or the motion of the moon around the earth, ignoring the effects of the sun—had already been solved by Newton. Hypnotized by Newton's epochal discoveries in dynamics, mathematicians expected to go ahead and solve the 3-body, and in general, the *n*-body problem. Poincaré's discovery of the homoclinic tangle was the first intimation of the profound truth, understood only in our times, of the chaotic character of almost all dynamical systems in dimensions higher than 2. They are unpredictable in the long run because unobservably small deviations in the data today eventually result in arbitrarily large differences in the outcome.

For a long time Poincaré's successors in the mathematical theory of dynamical systems turned away from this problem of the homoclinic tangle. Such a "strange attractor" cannot arise in two dimensions. In fact, dynamical systems in two dimensions of phase space are well understood, although of course many open questions remain, and it was generally expected that similar regularity could ultimately be proved in higher dimensions as well. Mathematicians tried in vain to prove that "most" dynamical systems are "well-behaved," that is, predictable in the distant future.

This was the project that Steve Smale of Berkeley undertook until a postcard from Norman Levinson of MIT called his attention to a paper of John Littlewood (1885–1977) and Mary Cartwright (1900–1998). Working on the Van der Pol equation of electrical engineering, they had found a kind of chaotic behavior similar to what Poincaré had discovered in the n-body problem. Smale reversed course and ultimately was able to show that in higher dimensions, unlike the plane, chaos rather than predictability is the norm.

The term "chaos" has been used to describe a wide variety of phenomena. It was introduced by Yorke and Li in relation to a different phenomenon.

Meanwhile, quite apart from these investigations in pure mathematics, the Hayashi group in Japan, particularly Yoshisuke Ueda, and Christian Mira and his group in Toulouse, France, were finding chaos by carrying out long calculations on analog computers.

These researchers were challenged by colleagues who ascribed their findings to "noise." They preferred to ignore the implications and questioned the focus of researchers who wandered so far from the results of traditional mathematical reasoning.

In a volume called *The Chaos Avant-Garde*, some of these researchers recall their frustrating experiences. Christian Mira wrote of the research on chaotic manifolds in Toulouse as follows: ". . . a point is worth to be mentioned about the local background of these researchers. The complex dynamics studies

Figure 2-9. Steve Smale, leading researcher in topology and dynamical systems. Courtesy of Archives of the Mathematisches Forschungsinstitut Oberwolfach.

of the Toulouse group have never occurred in a favorable environment. So my projects of developing this topic and applications were systematically refused during the 'prehistoric' times of 'chaos,' with the argument 'nobody is interested by such a matter,' which I frequently heard. . . . This in spite of the highly significant Smale's contribution of prime interest, leading in particular to the 'horseshoe' map properties (1963), and other rare interesting publications."[44]

Part of the problem was that many of the contributors to chaos theory came from applied fields, including engineering and physics. Another reason for the lack of acceptance of their findings was that some of the basic research was published in Russian journals which were not read by many western mathematicians.

A similar problem faced the Japanese researcher Yoshisuke Ueda. He was working in a laboratory where Professor Hayashi, the head of the team investigating electric oscillations, considered the unexpected chaotic results as inappropriate. He recommended that Ueda repeat the experiments "until the transient state settled to a more acceptable result."[45] In view of this opposition, Ueda waited until Hayashi was on sabbatical leave to continue his path-breaking research. But when he tried to publish some of his findings, they were rejected. He was caught in a difficult contradiction. While his work did not meet the standard of rigorous mathematical proof, it did correspond to phenomena that were observed in the real world. "While I wanted them to hear me out a little more sympathetically I also idolized mathematics . . . I had intentionally sent off the paper during Professor Hayashi's absence so I had a good excuse for not having it reviewed by him and it probably lacked certain fine editing because of it. But there was no way I could show the paper to Professor Hayashi since I knew he would make drastic changes and cut out what was essential to me. I couldn't compromise, however . . . I learned the hard way that changing the already established order in this world was a truly difficult task."[46]

Although it was hard to have this research accepted, eventually the work became known as "the Japanese attractor." Ueda refers to the early years of his research as "the dark hours before chaos was universally recognized."[47]

As more and more results emerged, confirming the prevalence of chaotic phenomena, Poincaré's discovery of the homoclinic tangle was remembered. In summarizing the long and complex 70-year history of the discovery of chaos, the American mathematician Ralph Abraham writes, "Triggered by mathematical discovery, The Chaos Revolution is a bifurcation event in the history of the sciences, comprised of sequential paradigm shifts in the various sciences. Perhaps it is also a major transformation in world cultural history: time will tell. Meanwhile, we are struck with the personal observation of the similarity in the sociological and psychological experiences of the various pioneers who have suffered from the novelty of their ideas, and the bravery of their convictions. We are deeply in their debt."[48]

Tenure Battle at Berkeley

Finally, here is a last example of tension in mathematical life. Divisions among mathematicians are not limited to their philosophical outlook or their attitude toward public awards and recognition. They also include issues of judgment about who should be part of highly prestigious mathematics departments. Decisions concerning membership involve the quality of an individual's research, collegiality, area of specialization, and possible hidden preferences concerning ethnicity, age, and gender. One of the best known cases is that of Jenny Harrison at the University of California, Berkeley. The central issue in the case was gender discrimination alleged by Harrison, who was denied tenure in 1986. After a long period of departmental and universitywide reviews, the university settled her case and granted her tenure and the title of full professor. Much has been written about this lengthy process, and some authors have said that the case opened a Pandora's box of issues about tenure, discrimination, and the law.[49] In addition to printed information, a recent personal interview with Harrison by John-Steiner provides a more complete picture of how she relied on inner resources during this conflict.

Harrison's earliest interest was nature. "I went to public school and I was basically self-taught because I wasn't getting anything out of school. And from the age of 5 until 15 I spent all of my time that was available to me outside of school in the woods." This attraction to the natural world has influenced Harrison throughout her life and can be seen in the way she explores mathematical landscapes.

In the interview she told John-Steiner about her visual view of the world and her enjoyment in exploring paths in the woods. An important influence in her life was her older brother, whose encouragement and sustained belief in Harrison's talent and determination contributed to her strength and self-confidence. "He was a great teacher, and got me interested in basic problems in physics when we were teenagers."

As a teenager she scored well in math contests, but she was primarily devoted to music. She thought that she was going to become a professional musician, but then she found she was uncomfortable performing in public. It has often been remarked that musical talent and mathematical talent seem to be associated. Of the people I have known personally, my graduate school office mate Leonard Sarason had earned a Master's degree at Yale under the composer Paul Hindemith. Richard Courant's eagerness to perform at the piano was notorious in his family. Courant's student Hans Lewy became a mathematics professor at Berkeley but could have been a concert pianist. The mathematical analyst Leonard Gilman of the University of Texas in Austin performed piano classics to entertain national meetings of the American Mathematical Society.

Once Harrison switched to mathematics at the University of Alabama, her work was quickly recognized, with an offer to become a Marshall scholar at the University of Warwick in England. She received excellent training at Warwick, but she encountered sexual harassment there. The issue of sex discrimination and male/female relationships in a discipline long dominated by men is intricately involved in the Harrison case. It affected her both at Warwick and at Berkeley.

Her dissertation research is highly regarded. She completed her thesis, titled "Unsmoothable Diffeomorphisms," in 1975. She worked as a postdoctoral fellow at the Institute for Advanced Study and as an instructor at Princeton University. She received a Miller Fellowship to Berkeley in 1977 and a subsequent offer for a tenure track assistant professorship at Berkeley. A year later, she accepted an offer from Oxford University for a position at Somerville College, where she spent 3 years while still retaining her Berkeley position. On returning to Berkeley in 1982, she announced that she had found a new and stronger counterexample to the Seifert conjecture. Harrison's work drew on new and delicate techniques from several fields and proved very difficult to write up for publication.[50] Once the manuscript was submitted, opposition by Michel Herman, a leading expert

Figure 2-10. Jenny Harrison, Berkeley professor of mathematics. Courtesy of Archives of the Mathematisches Forschungsinstitut Oberwolfach.

in the field, resulted in a more than usually cautious stance by the referees. Harrison further clarified what happened.

> Michel Herman was a highly regarded French mathematician who believed my methods were not only hopelessly wrong, but "dangerous" and did not hesitate to say these things to others. He feared my novel methods would be taken seriously by others and there would be no end of problems to correct. Besides that, he was working hard to find a counterexample to the Seifert Conjecture himself. Herman's opposition was taken seriously by referees who took years to accept the papers. They were convinced they would find a fatal flaw somewhere in the long and technical work, but finally relented and the papers were accepted in 1986. Herman had been wrong. In recent years, it has become apparent how important these papers were. They led the way to my broad extension of calculus, which unites the discrete to the smooth continuum, solving an old and important problem of mathematics that dates back to Newton and Leibniz.[51]

In 1986, after a many-stage process going up from a vote by the math department to the dean, two committees, and the chancellor, Harrison was denied tenure. She filed a formal complaint with Berkeley's Academic Senate Committee on Privilege and Tenure, which ruled against her. In her web page Harrison writes, "From the very beginning I only asked for a new and fair review, free from any question of gender bias."[52] Once denied, her next step was to file a lawsuit against the university in 1989 charging sex discrimination. The lawsuit was eventually settled by the university, based on the recommendation of seven distinguished members of the mathematics community chosen by the university. Their identities were kept strictly confidential, and they included two members of the math department who had not taken sides in the case. "If a majority of the review committee voted against me, I agreed to walk away, with no appeal. If a majority voted for me, I would be reinstated."[53] The committee reviewed Harrison's research subsequent to the original tenure decision and unanimously agreed that it was as good as the work of 10 other mathematicians who had received tenure in her department in the meantime. They recommended offering her a full professorship.

This decision at the university level was controversial in the math department. From the very beginning of this case there were deep divisions among the faculty members. One issue was, what are the criteria for tenure in an elite department such as Berkeley? Whatever these may be, the department's record in hiring female faculty members was considered by some faculty members to be very poor. As mentioned later in this book, Julia Robinson was appointed only after she became a member of the National Academy of Sciences. After Robinson's death, there was only one female member among 55 men in the department until Harrison's appointment.

Many faculty members believe that part of the university's willingness to appoint an independent committee and to accept their decision was their concern that this record of female absence would reflect poorly on Berkeley. Others argue that it was better for the dispute to be decided by knowledgeable academic experts rather than in a courtroom by a jury of 12 randomly selected citizens of Alameda County. Some members of the math department strongly supported Harrison. Her case was publicized by a Harrison support committee (chaired by Charity Hirsch, the wife of a well-known member of the department), including faculty from across the campus. It was financially supported by the legal funds of the American Association for University Professors and the American Association of University Women, who accepted Harrison's contention that the appeal mechanism within the university was inadequate for addressing her grievances.[54]

Once the decision was made by the university, the division within the department became even more pronounced. Some members objected to the department being overruled by the administration. Others were critical of Harrison's accomplishments. In a comprehensive article on this case written by Allyn Jackson, Dorothy Wallace said that the tenure process was quite subjective: "We are expected to be able to say that the candidate is at least as good as any of the people to whom we are comparing her or him. We are never asked to justify this judgment. In fact we are never asked to define what constitutes 'good' or 'as good as'. . . . I am driving at the fact that the process itself is wide open to any sort of individual or group prejudice that possibly could exist."[55] The opposition to Harrison involves only a few members of the department, but it has been vociferous and frequently hurtful to her.

During the years of fighting for her case, Jenny Harrison suffered from tonsil cancer but continued her fight without giving in to despair. In fact, her research continued to develop, and its originality and importance have been broadly recognized during the 12 years that have followed her reinstatement. Calvin Moore has written,

> During the period of appeals and litigation, Harrison had continued her research program, developing new lines of research

in geometric measure theory, and her work was supported by federal grants. After her reinstatement as professor, she has continued to develop her research program in geometric measure theory aimed at understanding multivariable calculus on domains with very irregular boundaries. In this she was building on and expanding ideas of Hassler Whitney. More recently, she has published work on a theory of domains based upon an extension of the Cartan exterior calculus to a normed space of domains that includes soap films, manifolds, rough domains, and discrete atomic structures. The analysis involved in her earlier work on the Seifert conjecture counterexamples with three minus epsilon derivatives had pointed the way to these more recent results. She has supervised the doctoral work of two students.[56]

Her case raises important issues about women's determination to make a place in a traditionally masculine culture. Harrison strongly believes in maintaining one's balance when confronted with difficult problems. She has tried to establish a balance between teaching, research, music, and love of nature. In this book, where we emphasize the emotional aspects of mathematics, this story from Berkeley reminds us how, even in a discipline that prizes objectivity and rationality, professional practice is not immune from long-held societywide assumptions about women's contributions to outstanding intellectual work.

The lived experience of mathematicians is influenced by the long history and the sense of awe in which their discipline is held. Members of the profession still address century-old problems, while constantly renewing their techniques and modes of thought. Recently, as mathematics has captured public interest and attention, the more hidden emotional aspects have come to the fore. Mathematicians rely deeply upon each other to evaluate their proofs and findings even as they compete for prizes and fame. This tension between community and competition has recently been made more complex by the greater heterogeneity of membership in what has traditionally been a rather exclusive group.

How have we answered our question, What is the culture of mathematics? We have tried to contribute to it, by examining the cognitive, emotional, and aesthetic aspects of the field and the social and competitive tensions. All these issues are, to some degree, found in other aspects of academic and scientific life, yet all have a special and definite flavor in the life of mathematicians.

Bibliography

Abraham, R. (2000). The chaos revolution: A personal view. In R. Abraham & Y. Ueda, (Eds.) *The chaos avant-garde: Memories of the early days of chaos theory*. Singapore: World Scientific, pp. 81–90.

Abraham, R., & Ueda, Y. (Eds.) (2000). *The chaos avant-garde: Memories of the early days of chaos theory*. Singapore: World Scientific.

Albers, D. J., & Alexanderson, G. L. (Eds.) (1985). *Mathematical people: Profiles and interviews*. Boston: Birkhäuser.

Atiyah, M. (1984). Interview. *Mathematical Intelligencer* 6(1), 9–19.

Barrow-Green, J. (1997). *Poincaré and the three body problem*. Providence, R.I.: American Mathematical Society.

Bell, E. T. (1965). *Men of mathematics*. New York: Simon and Schuster.

Boas, R. (1990). Interview. In D. J. Albers, G. L. Alexanderson & C. Reid (Eds.), *More mathematical people: Contemporary conversations*. Boston: Birkhauser, p. 25.

Brockman, J. (Ed.) (2004). *Curious minds*. New York: Pantheon Books.

Chang, Kenneth (2007). Journeys to the distant fields of prime. *New York Times*, March 13, 2007. Retrieved April 10, 2008 from http://www.nytimes.com/2007/03/13/science/13prof.html?_r=1& sq=The%20Mozart%20of%20MAth&st=nyt&oref=slogin&scp =1&pagewanted=print

Collins, M. A., & Amabile, T. M. (1998). Creativity and motivation. In R. J. Sternberg (Ed.), *Handbook of creativity*. Cambridge: Cambridge University Press, pp. 297–312.

Csikszentmihalyi, M. (1990). *Flow: The psychology of optimal experience*. New York: Harper Perennial.

da C. Andrade, E. N. (1954). *Sir Isaac Newton: His life and work*. Garden City, N.Y.: Anchor Books.

Davis, P. J., & Hersh, R. (1981). *The mathematical experience*. Boston: Birkhäuser.

Diacu, F., & Holmes, P. (1996). *Celestial encounters*. Princeton, N.J.: Princeton University Press.

Everett, D. (2005). Cultural constraints on grammar and cognition in Piraha. *Current Anthropology* (Aug./Sept.), pp. 621–646.

Gardner, H. (1993). *Creating minds*. New York: Basic Books.

Glas, E. (2006). Mathematics as objective knowledge and as human practice. In R. Hersh (Ed.). *18 unconventional essays on the nature of mathematics*. New York: Springer.

Hadamard, J. (1945). *The psychology of invention in the mathematical field*. New York: Dover.

Hálmos, P. R. (1985). *I want to be a mathematician*. New York: MAA Spectrum, Springer-Verlag.

Hardy, G. H. (1991). *A mathematician's apology*. Cambridge: Cambridge University Press.

Harrison, J. (2007). Web site.

Hersh, R. (Ed.). (2006). *18 unconventional essays on the nature of mathematics*. New York: Springer.

Jackson, A. (1994). Fighting for tenure: The Jenny Harrison case opens Pandora's box of issues about tenure, discrimination, and the law. *Notices of the American Mathematical Society 41*(3), 187–194.

John-Steiner, V. (2006). Harrison interview, December 4, 2006, Berkeley, Calif.

Krantz, S. G. (2002). *Mathematical apocrypha: Stories and anecdotes of mathematicians and the mathematical*. Washington D.C.: Mathematical Association of America.

Krantz, S. G. (2005). *Mathematical apocrypha redux: More stories and anecdotes of mathematicians and the mathematical.* Washington D.C.: Mathematical Association of America.

Macrae, Norman (1992). *John von Neumann.* New York: Random House.

Mandelbrot, B. (1985). Interview. In D. J. Albers & G. L. Alexanderson (Eds.). *Mathematical people: Profiles and interviews.* Boston: Birkhäuser.

Mira, C. (2000). I. Gumowski and a Toulouse research group in the "prehistoric" times of chaotic dynamics. In R. Abraham & Y. Ueda (Eds.). *The chaos avant-garde: Memories of the early days of chaos theory.* Singapore: World Scientific, pp. 95–197.

Moore, C. C. (2007). *Mathematics at Berkeley: A history.* Wellesley, Mass.: A. K. Peters.

Mozzochi, C. J. (2000). *The Fermat diary.* Providence, R.I.: American Mathematical Society.

Nasar, S. & Gruber, D. (2006). Manifold destiny. *The New Yorker,* August 28, 2006.

Osserman, R. (1995). *Poetry of the universe.* New York: Doubleday.

Parikh, C. (1991). *The unreal life of Oscar Zariski.* Boston: Academic Press.

Peter, R. (1990). Mathematics is beautiful. *Mathematical Intelligencer* 12 (1), 58–62.

Rota, G. C. (1997). *Indiscrete thoughts.* Boston: Birkhäuser.

Ruelle, D. (2007). *The mathematician's brain.* Princeton, N.J.: Princeton University Press.

Singh, S. (1998). *Fermat's last theorem.* London: Fourth Estate.

Schwartz, L. (2001). *A mathematician grappling with his century.* Basel: Birkhäuser.

Sternberg, R. J., & Lubart, T. I. (1991). An investment theory of creativity and its development. *Human Development, 34,* 1–31.

Ueda, Y. (2000a). Strange attractors and the origins of chaos. In R. Abraham & Y. Ueda (Eds.). *The chaos avant-garde: Memories of the early days of chaos theory.* Singapore: World Scientific, pp. 23–55.

Ueda, Y. (2000b). My encounter with chaos. In R. Abraham & Y. Ueda (Eds.). *The chaos avant-garde: Memories of the early days of chaos theory.* Singapore: World Scientific, pp. 57–64.

Ueda, Y. (2000c). Reflections on the origin of the broken-egg chaotic attractor. In R. Abraham & Y. Ueda (Eds.). *The chaos avant-garde: Memories of the early days of chaos theory.* Singapore: World Scientific, pp. 65–80.

Vygotsky, L. S. (1962). *Thought and language.* Cambridge, Mass.: MIT Press.

Weil, A. (1992). *The apprenticeship of a mathematician.* Basel: Birkhäuser.

White, L. A. (2006). The locus of mathematical reality: An anthropological footnote. In R. Hersh (Ed.). *18 unconventional essays on the nature of mathematics.* New York: Springer, pp. 304–319.

Wilder, R. L. (1981). *Mathematics as a cultural system.* Oxford: Pergamon Press.

Yau, S. T. (1998). *S. S. Chern: A great geometer of the twentieth century.* Singapore: International Press.

+ 3 +

Mathematics as Solace

When looking at mathematical life, we usually focus on its public face: the institutions in which mathematicians work, their interactions within their communities, their jokes and eccentricities, their prizes and competitions, their breakthrough discoveries. In the following pages we address the more personal consequences of loving mathematics. We ask, *"Is mathematics a safe hiding place from the miseries of the world?"*

According to Gian-Carlo Rota, "Of all escapes from reality, mathematics is the most successful ever. It is a fantasy that becomes all the more addictive because it works back to improve the same reality we are trying to evade. All other escapes—sex, drugs, hobbies, whatever—are ephemeral by comparison.... The mathematician becomes totally committed, a monster, like Nabokov's chess player who eventually sees all life as subordinate to the game of chess."[1]

While working on this book we were surprised at how many well-known mathematicians have created mathematics while in prison. We found five prisoners of war, in three different wars, plus two political prisoners and one convicted of evading military service. Alongside these, there was one who used mathematics to escape an excruciating toothache, one who was revived to life by a mathematical problem while bedridden and almost 90 years old, a novelist who was distracted by mathematics from his decades-long writer's block, and an idealistic youngster who was helped to endure the agony of participating in a senseless, brutal bombing war.

Absorption

In its mild form, escape is absorption. You know you're finally really getting into your problem when you dream about it every night. (No guarantee your dream will give you the solution!) They say Newton sometimes forgot both to eat and to sleep. To many, this is absent-mindedness. They say that Norbert Wiener, when walking down a corridor at MIT with a mathematical paper in his right hand, would come to an open classroom door, walk through the doorway and around the four walls of the classroom and then out again, guiding himself with his left hand against the wall, while still reading.

Blaise Pascal (1623–1662) was one of France's most illustrious sons, both in mathematics and in literature. He had renounced mathematics and science in favor of ascetic devotion to the Blessed Virgin, but he was still able to turn to mathematics in an emergency. Lying awake one night in 1658 tortured by a toothache, Pascal tried thinking furiously about the cycloid, hoping to take his mind off the excruciating pain. (The cycloid is the curve generated by a fixed point on a circle as the circle rolls along a horizontal track.) He was pleasantly surprised to notice that the pain had stopped. Pascal interpreted this as a sign from heaven: he was not sinning to think about the cycloid rather than about the salvation of his soul! He devoted 8 days to the geometry of the cycloid, solving many of the main problems concerning it.

John Littlewood, the famous collaborator of G. H. Hardy, got a new lease on life at the age of 89. After a bad fall in January 1975, he was taken to a nursing home in Cambridge and had very little interest in life. His colleague, Béla Bollobás, suggested the problem of "determining the best constant in Burkholder's weak L^2 inequality (an extension of an inequality he had worked on)." To Bollobás's immense relief and amazement, Littlewood became interested. Although he had never heard of martingales (which is the subject of the Burkholder inequality) he was eager to learn about them—at his age and in bad health! He was able

Figure 3-1. George Polya and John Littlewood. Source: *The Polya Picture Album: Encounters of a Mathematician.* Ed. G. L. Alexanderson. Boston: Birkhauser, 1985. Pg. 151. Reprinted by kind permission of Springer Science and Business Media.

to leave the nursing home a few weeks later. "From then on," Bollobás wrote, "he kept up his interest in the weak inequality and worked hard to find suitable constructions to complement an improved upper bound."[2]

The American novelist Henry Roth, author of *Call It Sleep*, lived for many years in a remote village in Maine, in the far northern United States. He was suffering from writer's block and attempted to help support his family by raising and slaughtering ducks and geese. To survive the Maine winters, he did calculus problems. In fact, he did *all* the problems in George B. Thomas' influential calculus text, and later visited Professor Thomas at MIT to tell about this feat.

Freeman Dyson is a famous physicist (who is nowadays also a regular contributor to the *New York Review of Books*.) In 1943 he was working 60 hours a week as a statistician for the Royal Air Force Bomber Command in the middle of a forest in

Buckinghamshire. He remembers it as a long, hard, grim winter. The bomber losses he was analyzing were steadily growing higher, and the end of the war was not in sight. Hardy knew that Dyson was interested in the Rogers-Ramanujan identities,[3] so he sent him a paper by W. N. Bailey that contained a new method of deriving identities of the Rogers-Ramanujan type. Dyson wrote, "In the evenings of that winter I kept myself sane by wandering in Ramanujan's garden, reading the letters I was receiving from Bailey, working through Bailey's ideas and discovering new Rogers-Ramanujan identities of my own. I found a lot of identities of the sort that Ramanujan would have enjoyed. My favorite was

$$\sum_{n=0}^{\infty} x^{n^2+n} \frac{(1+x+x^2)(1+x^2+x^4)\cdots(1+x^n+x^{2n})}{(1-x)(1-x^2)\cdots(1-x^{2x+1})} = \prod_{n=1}^{\infty} \frac{(1-x^{9n})}{(1-x^n)}.$$

In the cold, dark evening, while I was scribbling these beautiful identities amid the death and destruction of 1944, I felt close to Ramanujan. He had been scribbling even more beautiful identities amid the death and destruction of 1917."[4]

Another World War II story is about Olga Taussky-Todd and her husband, the numerical analyst John Todd. They were in London during the Blitz—the intensive bombing of the city by the German air force. The couple took advantage of these raids to get some work done while taking shelter on the ground floor of their apartment building. Todd told an interviewer that "During raids we wrote papers—about six in all—while the other twenty to thirty people chatted, slept, or read."

Prison Stories

Quite a few well-known mathematicians have served time as prisoners of war, from the Napoleonic War to World War II. At least two have been political prisoners—in the United States

Figure 3-2. Jean-Victor Poncelet. Courtesy of Smithsonian Institution Libraries, Washington, DC

and in Uruguay. An impressive amount of beautiful mathematics has in fact been created in prison, where it served to help the imprisoned mathematician survive his ordeal.

A major part of projective geometry was created in prison. In November 1812 Jean-Victor Poncelet (1788–1867), a young officer in the exhausted remnant of Napoleon's army retreating from Moscow under Marshal Ney, was left for dead on the frozen battlefield of Krasnoi. A Russian search party found him still breathing. In March 1813, after a 5-month march across the frozen plains, he entered prison at Saratov on the banks of the Volga. When "the splendid April sun restored his vitality," he commenced to reproduce as much as he could of the mathematics he had learned at the École Polytechnique, where he had been inspired by the new descriptive geometry of Monge and the elder Carnot. In September 1814 Poncelet returned to France, carrying with him the material of seven manuscript notebooks written at Saratov. Bell writes that this work "started a tremendous surge forward in projective geometry, modern synthetic geometry, geometry generally, and the geometric interpretation of imaginary numbers that present themselves in geometric manipulations, as ideal elements of space."

Leopold Vietoris, the Austrian topologist who died in 2002 at the age of 111, served as a mountain guide for the Austro-Hungarian army in World War I while working on his thesis, "To Create a Geometrical Notion of Manifold with Topological Means." (A "manifold" is a generalization to higher dimensions of two-dimensional smooth curved surfaces such as spheres or cylinders.) Just before the armistice, on November 4, 1918, Vietoris was captured by the Italians. He completed his thesis while a prisoner of war.

Two other mathematicians in the Austro-Hungarian army in World War I were taken prisoner, not by Italians but by Russians. Eduard Helly of Vienna and Tibor Radó of Budapest met in a prison camp near Tobolsk in 1918. Radó had just begun university when he enlisted as a lieutenant and was sent to the Russian front. Helly was already a research mathematician; he had proved the Hahn-Banach theorem in 1912 before either Hahn or Banach. (This theorem is an essential tool in functional analysis. It permits one to extend a linear functional from a subspace to the whole space without increasing its magnitude.) Radó had studied civil engineering at the University of Budapest. In the Russian prisoner-of-war camp, Helly became Radó's teacher. Radó escaped from the prison camp and made his way north to the arctic regions of Russia. There indigenous arctic dwellers befriended him and gave him hospitality. He slowly trekked westward for thousands of miles, reaching Hungary in 1920. It had been 5 years since he had left the university in Budapest. Helly had shown him the fascination of research in mathematics and so he went to the University of Szeged to study with the Hungarian analyst Frigyes Riesz. He assisted Riesz in his great book on functional analysis and in

1929 migrated to the United States. He founded the graduate program in mathematics at Ohio State University and was a leading authority on the theory of surface measure.

Being a mathematician could be a serious detriment in prison. The French analyst and applied mathematician Jean Leray was a German prisoner of war for 5 years in World War II. If the Germans had known about his competence in fluid dynamics and mechanics, they might have tried to force him to work for them. So he turned his minor interest in topology into a major one and while in prison did research only in topology. In fact, he created sheaf theory, which soon became one of the principal tools in algebraic topology. (We give some details about sheaf theory in the next chapter in writing about Alexandre Grothendieck.) Nevertheless, once he was free, Leray returned to analysis, leaving topology to others.

The South African statistician J. E. Kerrich was visiting Denmark when the Nazis invaded in 1940. The Danes saved the British citizens who were then in their country by agreeing to intern them so that they would not be taken to Germany. Kerrich made use of his time in confinement by tossing a coin *10,000 times* and recording the results. He then wrote a little textbook, *An Experimental Introduction to the Theory of Probability*, based on analyzing the data from his experiment.

The French number theorist André Weil, like his compatriots Poncelet and Leray, had a spectacularly productive time in prison. In the summer of 1939, war with Germany was imminent, and Weil was under orders for military service. "This was a fate that I thought it my duty, or rather my dharma, to avoid," he wrote in his autobiography. He departed for Finland. By bad luck, the Russians invaded Finland a few months later. "My myopic squint and my obviously foreign clothing called attention to me. The police conducted a search of my apartment. They found several rolls of stenotypewritten paper at the bottom of a closet. . . . There was also a letter in Russian, from Pontryagin." After 3 days in prison, he was unexpectedly released at the Swedish border.[5] Shipped back to France by way of

Sweden and Scotland, Weil spent 3 months in jail in Rouen. His friend Henri Cartan wrote to him, "We're not all lucky enough to sit and work undisturbed like you." On April 7, 1940, he wrote to his wife Eveline, "My mathematics work is proceeding beyond my wildest hopes, and I am even a bit worried—if it's only in prison that I work so well, will I have to arrange to spend two or three months locked up every year?" On April 22, he wrote her, "My mathematical fevers have abated . . . before I can go any further it is incumbent upon me to work out the details of my proofs. . . ." On May 3, 1940, he was sentenced to 5 years in prison, which was immediately commuted if he agreed to serve in combat. On June 17, 1940, "the command came to abandon our machine guns and join our regiment on the beach. We were boarded on a small steamship . . . the next morning we were in Plymouth."[6] Weil eventually reached the United States to continue his illustrious career.

Mathematics and Politics

Chandler Davis, the editor of *The Mathematical Intelligencer*, was a schoolmate of one of the authors, a math grad student at Harvard when R. H. was an undergraduate English major. During the 1950s McCarthyite red scare, Davis's career was interrupted when he refused to answer questions asked by the U.S. House of Representatives Committee on Un-American Activities. He proudly referred to his revolutionary ancestry—in the *American* Revolution—and refused to cooperate in proceedings that violated the First Amendment to the U.S. Constitution guaranteeing freedom of speech. Davis was fired by the University of Michigan from his job as assistant professor of mathematics. He was convicted of contempt of Congress, and after exhausting appeals he was confined for 6 months in the federal penitentiary in Danbury, Connecticut. Then, when he was released, he was totally blacklisted by universities in the United States. The great Canadian geometer Donald Coxeter invited him to apply to

Figure 3-3. Chandler Davis, mathematician at the University of Toronto. Courtesy of Sylvia Wiegand.

the University of Toronto. At first the government refused Davis entry, but after a letter-writing campaign they relented. He moved to Canada to teach at the University of Toronto. A 1994 special issue of *Linear Algebra and Its Applications* celebrating his contributions to matrix theory describes his time in prison: "Throughout this ordeal, Chandler maintained his research interest in mathematics. He also maintained his sense of humor. A footnote in his paper on an extremal problem, conceived while he was in prison but published afterward, reads: 'Research supported in part by the Federal Prison System. Opinions expressed in this paper are the author's and are not necessarily those of the Bureau of Prisons.'" Davis says that this "elegant phrasing" was actually suggested to him by Peter Lax, a long-term member of the Courant Institute, whom he thanks.

Let it be understood that not every imprisoned mathematician fared as well as André Weil and Chandler Davis. The Uruguayan analyst José Luis Massera writes that he was drawn into political activity under the influence of refugee mathematicians from Fascism, Luis Santalo from Spain and Beppo Levi from Italy. Massera has left us a detailed account, both of his career before prison and of his time in prison. "That epoch, around the great movement of solidarity with the Spanish people, I began the political activity which I shared with mathematics for the rest of my life."[7] In 1943 he became a Communist.

After completing his degrees in Montevideo, Uruguay, Massera won a Rockefeller grant to go to Stanford where he worked with Polya and Szegö. But then he became more interested in differential equations (equations involving the rate of change of the unknown function), so he transferred to the East Coast, where he commuted between New York and Princeton, working simultaneously with Richard Courant on minimal surfaces and with Solomon Lefschetz on topological methods for ordinary differential equations. In 1966, after returning to Uruguay, he published his well-known book *Linear Differential Equations and Function Spaces* with his student J. J. Schaffer. He was also elected as a Communist Deputy to the Parliament of Uruguay. When the military dictatorship outlawed the party in 1973, he became its underground director.

He was imprisoned in October of 1975. On the first day, while he was standing at attention with hands and ankles tied, a soldier took him by the shoulder and shoved him. He fell and fractured his right leg. "Despite which," he later wrote, "the drill continued until they were convinced that I couldn't stand. They laid me on a wire cot where I remained for a month without help." He was finally taken to a military hospital where he was X-rayed and given a cane. "My own body was required to mend the fracture." (In fact, for the rest of his life Massera walked with one leg shorter than the other.)

He remained imprisoned for 9½ years in a prison known as the Penal de Libertad; the ironic name was due to its location on the outskirts of a town of that name. The cells of the prison housed two inmates each. They were allowed 1 hour a day of recreation.

Figure 3-4. José Luis Massera of Uruguay and Lee Lorch of Canada, mathematicians and fighters for human justice. Soruce: Archivo General de la Universidad de la República (Uruguay), Subfondo Archivos Personales, José Luis Massera, Box 3 Folder 3B.

Then we could engage in sports, and those who didn't, like in my case, walked conversing with another prisoner. Human relationships were almost exclusively limited to the cell. . . . In it one could read books of the good library that had been formed with donations that families of prisoners had made when it was permitted. Of course, political books were excluded . . . as well as mathematics books. Who knows what mysterious messages could be conveyed by those odd and incomprehensible symbols?

My cellmates were various including Communists and Tupamaros, with whom I conversed freely on the most varied topics. [Tupamaros were a movement of urban guerrillas, a tactic rejected by the Communists.] A paper factory worker, a Communist, was with me for years and we became great friends; he was very intelligent and restless, we talked on the most diverse themes. I could give him little courses on physics, chemistry, etc., which he absorbed with passion. In other cells there were prisoners who were young mathematicians like Markarian and

Accinelli, whom I saw only during the recreations and collective tasks; running some risks we produced some small mathematical works like one entitled "Is it true that two plus two is always four?" which might interest and intrigue the non-mathematician prisoners.

During all this, Martha, my wife had also been imprisoned, she was tortured and interned in a women's jail, in what was formerly a monastery. She was there for three years until some time during the year 1979; she was able to recover our apartment, which had been occupied and sacked by the military.

We can add to Massera's own memoir a note called "*Recuerdos*" (Memories), part of an article written by Elvio Accinelli and Roberto Markarian, mathematicians who shared with Massera more than 3 years of imprisonment.

Written in secret, with tiny handwriting, manuscripts were carried from cell to cell by the prison inmate who delivered bread or tools, who risked with this audacity being punished and sent to the "Isla" [isolation].[9] Those little papers circulated in open defiance and Massera wrote about dialectic, logic and mathematics, making real our affirmation that science and culture cannot be destroyed. At that time and place, to think was entirely prohibited. To demonstrate by some means what was thought, was an act of defiance and bravery, beyond the intrinsic value that the written or demonstrated material might have.

And in those conditions we conquered a space to think and discuss. . . . It was as if, despite everything, in the interstices of repression one lived freely.

One day, a book on *Hilbert's Problems*, which had been sent and dedicated to Massera by Lipman Bers (then President of the American Mathematical Society), came to be seen by some prison inmates. Who knows how much pressure on the part of many international organizations allowed Massera to receive such a magnificent present in his cell.[10] And who knows how much compassion there was in the prison guard that permitted that such a book be found in the prison wing.

To be a Communist was dangerous. And in that place it was also dangerous to be a mathematician. . . . Life and mathematics kept those of us who were the protagonists of this story united. Massera was for us a teacher, beyond the strictly scientific arena. And a friend with whom we shared with pride many joyous moments as well as other kinds.

Today we scarcely write or speak of these things. Nevertheless we remember those years, that some consider "empty," often with a smile on our faces. But we forget neither the pain nor the learning of life that we experienced. It is strange to say, but those years of imprisonment and isolation had positive aspects for us. We wouldn't be who we are, including professionally, without the "stain" of that period.

An international campaign of protest on Massera's behalf was carried on for years, led by Laurent Schwartz in France, by Lipman Bers and Chandler Davis in the United States, and by Lee Lorch and Israel Halperin in Canada. On March 3, 1984, Massera was set free. "For months, my house was invaded by hundreds of friends who came to greet me."

My Thoughts Are Free

Mathematicians are not the only prisoners who found in their profession comfort and solace. In a fascinating collection entitled *The Great Prisoners*, Isidore Abramowitz collected the letters and testimonies of men and women starting with Socrates and ending with prisoners from the 20th century. Many of these documents present deeply felt justifications for the writers, philosophers, and politicians who have been unjustly incarcerated. Others, such as Fyodor Dostoevsky, write of their experience while in jail to members of their families or the public. The similarity between the mathematicians and the men and women included in this volume is that in all these groups, when the prisoners had access to pencil and paper, they could find ways to escape from their dire circumstances.

The English writer Oscar Wilde had courted sensationalism in his writings and behavior but paid a wrenching price for his homosexuality. When accused of seduction by the father of one of his lovers, he was brought to trial, lost the case, and spent two very difficult years in prison. In a letter to a friend he wrote:

> I need not remind you that mere expression is to an artist the supreme and only mode of life. It is by utterance that we live. Of the many, many things for which I have to thank the Governor there is none for which I am more grateful than for his permission to write fully and at as great a length as I desire. For nearly two years I had within a growing burden of bitterness, of much of which I have now got rid. On the other side of the prison wall there are some poor black soot-besmirched trees that are just breaking out into buds of an almost shrill green. I know quite well what they are going through. They are finding expression.[11]

These diverse prisoners share the ability to sustain their intense inner life even under the most horrendous circumstances. When circumstances allow them to draw upon their passion and knowledge, it can become a means for survival.

Let's end with an uplifting story. In his youth Pál Turán, a member of Erdős's Anonymous Group (chapter 6) was imprisoned in a Fascist labor camp, from which he maintained a mathematical correspondence with Erdős. The Hungarian labor forces into which Turán was drafted were formed to support the army's operations. They consisted of people considered too untrustworthy to be given arms in the regular army—political opponents, Gypsies, and Jews. The young men who made up these forces were unarmed, and they served under regular army officers. They were ordered to clear railway lines and build stag-

Figure 3-5. Pál Turán and Vera Sós, a Hungarian couple. He was a number theorist, she is a combinatorialist. Courtesy of Archives of the Mathematisches Forschungsinstitut Oberwolfach.

ing areas close to combat zones. When war came, they were helpless when attacked because they had no weapons. They perished in great numbers. For Turán, it was a time of great pain, but he continued to work, using mathematical problem solving as his refuge.

In September 1940 I was called for the first time to serve in a labor camp. We were taken to Transylvania to work on building railways. Our main work was carrying railroad ties. It was not very difficult work, but any spectator would have recognized that most of us did it rather awkwardly. I was no exception. Once, one of my more expert comrades said so explicitly even mentioning my name. An officer was standing nearby, watching us work. When he heard my name, he asked the comrade whether I was a mathematician. It turned out that the officer, Joseph Winkler, was an engineer. In his youth he had placed in a mathematical competition; in civilian life he was a proofreader at the print shop where the periodical of the Third Class of the Academy (Mathematical and Natural Sciences) was printed. There he had seen some of my manuscripts. All he could do for me was to assign me to a wood-yard where big logs for railroad buildings were stored and sorted by thickness. My task was to show incoming groups where to find logs of a desired size. This was not so bad. I was walking outside all day long, in the nice scenery and the unpolluted air. The [mathematical] problems I had worked on in August came back to my mind, but I could not use paper to check my ideas. Then the formal extremal problem occurred to me, and I immediately felt that this was the problem appropriate to my circumstances. I cannot properly describe my feelings during the next few days. The pleasure of dealing with a quite unusual type of problem, the beauty of it, the gradual approach of the solution, and finally the complete solution made these days really ecstatic. The feeling of intellectual freedom and of being, to a certain extent, spiritually free of oppression only added to this ecstasy.[12]

After Turán's cry of ecstasy, any more words may seem an anticlimax. But we will conclude by pointing out the advantage imprisoned mathematicians have over other scientists. Poncelet needed only pencil, paper, and his memories of the Polytechnique to lose himself in projective geometry. Turán, without even pencil or paper, was able to re-create his world of combinatorial identities and estimates. According to Vladimir Arnold, mathematics is just "the part of physics where experiments are cheap." No need for a lab or even for a library, just your mind and its contents!

Bibliography

Abramowitz, Isidore (1946). *The great prisoners*. New York: E. P. Dutton.

Albers, Don. (2007). John Todd—Numerical mathematics pioneer. *College Mathematics Journal* 38(1), 11.

Elvio Accinelli, & Markarian, Roberto (1996). *Recuerdos* (Memories). In *Integrando* (Integrating). Translated by Frank Wimberly. *Centro de Estudiantes de Ingenieria* (Center of Engineering Students).

Bell, E.T. *Men of mathematics* (from Poncelet's Introduction to his *Applications d'analyse et de geometrie*, 1822, 1862).

Ballobás, Béla. (Ed.) (1986). *Littlewood's miscellany*. Cambridge: Cambridge University Press.

Choi, M.-D., & Rosenthal, P. (1994). A survey of Chandler Davis, *Linear Algebra and Its Applications 208/209*, 3–18.

Dyson, F. J. (1988). A walk through Ramanujan's garden. In G. E. Andrews et al. *Ramanujan Revisited*. Boston: Academic Press, pp. 7–28.

Helly, Eduard. MacTutor on-line mathematics biography.

Kerrich, J. E. (1946). *An experimental introduction to the theory of probability*. Copenhagen: Einar Munksgaard.

Leray, Jean (2000). *Notices of the American Mathematical Society 47*(3), 350–359.

Massera, J. L. (1998). *Recuerdos de mi vida academica y política* (Memories of my academic and political life). Lecture delivered at the National Anthropology Museum of Mexico City, March 6, 1998, and published in *Jose Luis Massera: The scientist and the man*. Montevideo, Uruguay: Faculty of Engineering. Translated by Frank Wimberly.

Massera, J. L., & Schaffer, J. J. (1966). *Linear differential equations and function spaces*. New York: Academic Press.

Radó, Tibor. MacTutor on-line mathematics biography.

Reitberger, H. (2002). Leopold Vietoris (1891–2002). *Notices of the American Mathematical Society 49*(10), 1232.

Roberts, S. (2006). *King of infinite space*. New York: Walker and Company.

Rota, G. C. (1990). "The lost café," in *Indiscrete thoughts*. Boston: Birkhäuser.

Schwartz, L. (2001). *A Mathematician grappling with his century*. Boston: Birkhäuser.

Sigmund, A. M., Michor, P., & Sigmund, K. (2005). Leray in Edelbach. *Mathematical Intelligencer 27*(2), 41–50.

Turán, P. (1997). Note of welcome. *Journal of Graph Theory 1*(1), 1.

Weil, A. (1992). *The apprenticeship of a mathematician*. Boston: Birkhäuser.

+4+

Mathematics as an Addiction: Following Logic to the End

The question is sometimes asked, To be a great mathematician, does being crazy help? The simple and straightforward answer is, No, of course not. Working in a university math department, or attending the meetings of the American Mathematical Society, one cannot help observing the pervasive normality. Still, there is something different about mathematicians compared to, say, chemists or geologists or even English professors. It is possible to be "crazy"—that is, conspicuously eccentric, very odd, even antisocial—and still hold a job as a math professor. Even, perhaps, as an industrial mathematician in certain organizations. If you are really good at solving hard problems and can communicate with other human beings well enough to convince them of what you have done, then in many universities no one will care too much if you work all night and sleep till noon or are careless about keeping your hair combed and your shoelaces tied. There is a certain casualness or "sloppiness" detectable at math meetings that would not be present at a convention of brain surgeons or chemical engineers. Standards of conformity and conventionality are more liberal and tolerant in the mathematical community than in many other academic or professional communities.

Of course, it can still happen that a mathematician succumbs to real mental illness. The case of John Nash became famous through Sylvia Nasar's book, *A Beautiful Mind*, which even became a Hollywood movie. After a brilliant early career, Nash was disabled by paranoid schizophrenia. He ultimately made a good recovery, in time to receive a Nobel prize. It would be very complicated to disentangle the relation, if any, between his mathematical genius and his madness.

As we mentioned in the last chapter, some mathematicians under extreme stress have found that their discipline provided solace and relief. Others, lacking a sense of balance, have put themselves at risk by making mathematics the only focus of their lives. *For some mathematicians, it seems, mathematics can be a destructive addiction.* In some cases, there was a specifically mathematical style or flavor to their delusions.

Our first example is by far the most important and takes up about half of this chapter. Alexandre Grothendieck was one of the preeminent mathematicians of his time, most famous for completely re-creating and transforming algebraic geometry. His life story is truly fascinating and amazing, raising questions not only about the potentially addictive nature of mathematics but also about the psychological destruction resulting from the cataclysmic wars and persecutions in Europe in the 20th century. In contrast with most mathematicians, Grothendieck wrote at length about his emotions. This apparent need to share his feelings and experience, and to describe the sources and direction of his discoveries, may be rooted in his lonely childhood.

After describing his achievements and his descent into solitary delusions, we present five cases of actual criminal or suicidal insanity in other mathematicians, sometimes with claims that their violent actions were "logical." Of course, none of the other mathematicians we discuss in this chapter are comparable to Grothendieck as creators of great mathematics. The lesson is that mathematics can indeed be dangerous for vulnerable minds in unfavorable circumstances.

We now turn to the amazing life story of Alexandre Grothendieck, from his heritage of revolutionary rebellion to his tragic final alienation from the mathematical community. Today Alexandre Grothendieck is a pacifist hermit, living at a secret address in a remote village in the Pyrenees Mountains. From 1950 to 1970, Grothendieck reshaped functional analysis and algebraic geometry. In 1970 he started his withdrawal from the famous institutions of the mathematical world, declaring them corrupt and vicious, although he continued to create some mathematics for another 17 years.[1]

Understanding Grothendieck's creations requires serious preparation in algebraic geometry and category theory. We can only offer partial and incomplete renderings. We get help from his famous or infamous 1000-page unpublished work *Récoltes et Semailles (Reaping and Sowing),* as partially translated by Roy Lisker,[2] in which he combines personal and mathematical reflections. As Allyn Jackson says, it is "a dense, multi-layered work that reveals a great and sometimes terrifying mind carrying out the difficult work of trying to understand itself and the world. . . . He often succeeds at describing things that at first glance would seem quite ineffable."[3] Jackson's brilliant biography and tribute has contributed to our own presentation, together with materials from Leila Schneps and the Grothendieck circle.

Early Years

Alexandre Grothendieck's father, Alexander (Sascha) Shapiro, was born around 1890 to Hasidic Jewish parents in Novozybkov, a little town near the place where Russia, Ukraine, and Belarus now meet. At age 17 he was arrested for joining in the unsuccessful 1905 revolution in Russia, but his youth saved him from a death sentence. He spent 10 years in prison in Siberia after running away and being recaptured a few times. He lost an arm, attempting suicide to avoid capture by the police. Upon release in 1917 he quickly became a leader of the Socialist-Revolutionaries of the Left, a party that was soon outlawed by Lenin. He then took part in several other failed European revolutions. During the 1920s he participated in the armed clashes of the leftist parties opposing Hitler and the Nazis in Germany. He supported himself as a street photographer. In Berlin he met Hanka (Johanna Grothendieck). She had become part of the radical avant garde after escaping from her Lutheran bourgeois family in Hamburg. On March 28, 1928, Hanka had a son, Alexandre.

When Hitler took power in 1933, Sascha fled to Paris, and Hanka soon followed him. She left 5-year-old Alexandre hidden near Hamburg in a libertarian private school run by a Christian idealist, Wilhelm Heydorn. Dagmar, Heydorn's wife, later recalled young Alexandre as very free, completely honest, and lacking in inhibitions. As he recalled in *Récoltes et Semailles*, "When I was a child I loved going to school. I don't recall ever being bored at school. There was the magic of numbers and the magic of words, signs and sounds. And the magic of rhyme. In rhyming there appeared to be a mystery that went beyond the words. . . . For a while everything I said was in rhyme. . . . I wasn't what would be considered 'brilliant.' I became thoroughly absorbed in whatever interested me, to the detriment of all else, without concerning myself with winning the appreciation of the teacher."

The sudden separation from his parents was very traumatic for him. They were apart for 6 years. In 1939 war was imminent and political pressure was increasing. The Heydorns could no longer keep their foster children. Grothendieck was an especially difficult case—he looked Jewish. Through the French consulate in Hamburg, Dagmar managed to reach Shapiro in Paris and Hanka in Nimes. In May 1939, 11-year-old Alexandre was put on a train traveling from Hamburg to Paris.

Sascha and Hanka had gone to Spain in 1936. There Sascha fought with the anarchist militia in the civil war against Franco's Fascists. After their defeat, Sascha and other loyalist fighters fled to France. In 1939 Alexandre spent a brief time with both his parents before his father was taken away and imprisoned in Le Vernet, the worst of the French detention camps. In October

1940, after France capitulated to Hitler, the Vichy government shipped Shapiro and others off to be killed at Auschwitz. Alexandre and his mother were left to survive as best they could.

In his recollections Grothendieck wrote:

> For my first year of schooling in France, 1940, I was interned with my mother in a concentration camp, at Rieucros, near Mende. [While in this camp, Hanka contracted tuberculosis, from which she continued to suffer until she perished of it in 1957.] It was wartime and we were foreigners—"undesirables" as they put it. But the camp administration looked the other way when it came to the children in the camp, undesirable or not. . . . I was the oldest and the only one enrolled in school. It was 4 or 5 kilometers away, and I went in rain, wind and snow, in shoes if I was lucky to find them, that filled up with water. . . . During the final years of the war, during which my mother remained interned, I was placed in an orphanage run by the "*Secours Suisse*," at Chambon sur Lignon. Most of us were Jews, and when we were warned (by the local police) that the Gestapo was doing a round-up, we all went into the woods to hide for one or two nights. . . . This region of the Cevennes abounded with Jews in hiding. That so many survived is due to the solidarity of the local population.[4]

At the College Cevenol, Alexandre was remembered as being very intelligent and always immersed in thought, reading, and writing. He was a fierce chess player, loud, nervous, and brusque. In his recollections he recalled that he devoured his textbooks as soon as he got them, hoping that in the coming year he would really learn something interesting. He writes, ". . . I can still recall the first 'mathematics essay.' The teacher gave it a bad mark. It was to be a proof of 'three cases in which triangles were congruent.' My proof wasn't the official one in the textbook he followed religiously. All the same, I already knew that my proof was neither more nor less convincing than the one in the book, and was in accord with the traditional spirit of 'sliding this figure over that one.' It was evident that this man was unable or unwilling to think for himself in judging the worth of a train of reasoning."[5]

Once the war ended in May 1945, he and his mother lived outside Montpellier. There he attended the university, which offered him little but time, during which he developed some of his own ideas and unknowingly rediscovered some classical mathematics. In the autumn of 1948 he went to Paris, carrying a letter of recommendation to Henri Cartan from his calculus teacher at Montpellier. Cartan's seminar attracted the most brilliant and aggressive young mathematicians, graduates of the elite École Normale Superieure. Although Grothendieck was an unknown foreigner and a newcomer from the provinces, he was accepted among them, and his talent was recognized. In fact, several of the young men he met then would later become his collaborators. Nevertheless, in view of the deficiencies in his preparation at Montpellier, it seemed that he would do better in a less pressured environment; so he was advised to go to Nancy to study under Laurent Schwartz. To Laurent Schwartz, working in a somewhat isolated academic environment, "the most fantastic gift to Nancy came in the person of Alexandre Grothendieck."[6]

Schwartz and his wife, Helene, gave their new colleague a sense of belonging seldom experienced by a young man who had often lived as an outsider. They were caring people whose interests went beyond mathematics. Here we insert, parenthetically, Schwartz's political history. In Schwartz's younger days he was an active Trotskyist. In 1943, in hiding both as a Jew and as a Trotskyist, "life had become a constant series of dangers. You had to keep your eyes open and stay lucid. For this reason I decided to stop mathematical research during this period. I kept myself from the distraction of research which might have led me to relax my vigilance." For the rest of his life Schwartz was a leading activist for human rights. He opposed both of France's colonial wars, in Algeria and Indo-China, as well as Stalinist and fascist repression. In revenge for opposing the war in Algeria, his apartment building was plastic-bombed by an underground pro-war group, the Secret Army Organization (OAS).[7]

As Schwartz's student, Grothendieck wrote a dissertation of over 300 pages containing a vast generalization and abstraction of Schwartz's famous theory of distributions. Schwartz called it "a masterpiece of immense value." He wrote, "It was difficult. I spent six months full time on it. What a job, but what a pleasure!"[8] At schools dominated by Bourbaki, it produced a major transformation of functional analysis, although specialists in more application-oriented places like the Courant Institute continued to concentrate on the more traditional Banach space and Hilbert space. Nevertheless, because Grothendieck had no citizenship and as a matter of principle refused to apply for citizenship he couldn't be hired at a French university. He spent a few years in Paris as a postdoctoral fellow. Then he taught in São Paulo, Brazil. He spent part of 1955 in the United States in Kansas and at Chicago. Then he returned to France, supported by fellowships and without a regular job. That was when he changed his mathematical interests from analysis to geometry. He wrote, "It was as if I'd fled the harsh arid steppes to find myself suddenly transported to a kind of 'promised land' of superabundant richness, multiplying out to infinity wherever I placed my hand on it, either to search or to gather. . . . "[9]

In 1957 there was established, first in Paris and then on a hill in the countryside southwest of Paris, the French counterpart of Princeton's Institute for Advanced Study, the Institut des Hautes Études Scientifiques (IHES). Grothendieck and Jean Dieudonné were chosen as the two math professors. Since the IHES is not part of the French government, Grothendieck's statelessness was not a problem.

His mother Hanka had been nearly bedridden for several years, suffering from tuberculosis and severe depression. She and Alexandre had grown inseparable. In his personal writings the image of the mother appears repeatedly as the source of life and creativity. In her last months she was so bitter that his life became extremely difficult. A close friend of hers named Mireille, several years older than Alexandre, helped him care for Hanka during her last months. Mireille was fascinated and overwhelmed by Alexandre's powerful personality and fell in love with him.

Hanka died in December 1957. Her death so shocked Grothendieck that he left mathematics for a few months. But then he returned to mathematical research and married Mireille.

In 1958 Oscar Zariski invited Grothendieck to visit him at Harvard. But in order to get a U.S. visa, he had to sign an oath pledging not to work to overthrow the U.S. government. This Grothendieck refused to do. Zariski wrote to him, warning that Grothendieck's views might land him in prison. Grothendieck responded that that would be fine, as long as he could have books and visits from students. In February 1959 Mireille bore their first child, Johanna.

In March 1959 Grothendieck started giving seminars on algebraic geometry at the IHES. In collaboration with Jean-Paul Serre, with whom he shared many of his emergent ideas in correspondence as well as in person, he was building on ideas put forward by Jean Leray and André Weil. It is said that he worked 12 hours a day, 365 days a year, for 10 years. In his office at the IHES hung a portrait of his father, Sascha Shapiro. It was the only decoration. In 1961 he did visit Harvard, with Mireille as his wife. In July a son named Alexander was born, called Sasha after Alexandre's father.

During these years of deep absorption in mathematics, Grothendieck paid little attention to the larger world and politics. He seldom read the newspapers. But after the outbreak of the Algerian uprising against France, he changed. On October 5, 1961, there was a curfew on French Muslims from Algeria. On October 17 thousands of Algerians took to the streets of Paris to protest. There was a police massacre that day that left dozens of bloody bodies piled in the streets or floating down the Seine.

During the war against Algeria, Grothendieck refused to demand special exemption from military service for mathematics students. Rather, he wrote to Serre, "The more people there are who, by whatever means, be it conscientious objection, desertion, fraud or even knowing the right people, manage to extricate themselves from this idiocy, the better."[10]

In 1965 Mireille and Alexandre's third child was born: a son, Mathieu. In this period, Mireille described him as working all

Figure 4-1. Alexandre Grothendieck (center) among his colleagues. Courtesy of Archives of the Mathematisches Forschungsinstitut Oberwolfach.

night by the light of a desk lamp. She slept on a sofa in the study near him. Occasionally she would wake to see him slapping his head with his hand, trying to get ideas out faster.

Grothendieck poured out new ideas, and Dieudonné kept up as his scribe. Grothendieck was author or coauthor of around 30 volumes in the IHES blue series, most of which comprise over 150 pages. *Éléments de Géométrie Algébrique* (EGA), which he created in collaboration with Jean Dieudonné, and *Séminaire de Géométrie Algébrique* (SGA), consist of about 10,000 pages. His other works add a couple of thousand more pages. It was too much for Grothendieck to write, so he depended on a large group of more or less willing and able students and colleagues. There was a sense that a revolution was underway, as his ideas transformed algebraic geometry into one of the most abstract and technical fields in mathematics.

Grothendieck's Theories

In the next few paragraphs we will try to give an idea of this work accessible to the curious nonmathematician. We quote extensively from *Récoltes et Semailles*. Some of these quotes are vague and hard to comprehend precisely. But they are Grothendieck's own words! Nowhere else in the literature have we seen any account of these ideas written for the nonprofessional. Some readers may wish to skip over these paragraphs.

To begin savoring a slight taste of the flavor of this mathematics, consider first, Why did mathematicians need negative numbers? Because we wanted to be able to subtract any integer from any other regardless of which is bigger. Then we needed complex numbers because the solutions of quadratic and cubic equations cannot be fully understood within the domain of real numbers. And next we needed things called quaternions (or vectors), in order to use algebraic methods in three-dimensional space. At each stage, new kinds of numbers were created to solve more difficult problems.

Similarly, three daring conjectures by André Weil demanded the creation of new kinds of geometries. Grothendieck wrote:

> These utterly astounding conjectures allowed one to envisage, for the new discrete varieties (or "spaces"), the possibility for certain kinds of constructions and arguments which to that moment did not appear to be conceivable outside of the framework of the only "spaces" considered worthy of attention by analysts—that is to say the so-called "topological" spaces (in which the notion of continuous variation is applicable). One can say that the new geometry is, above all else, a synthesis between these two worlds, which, though next-door neighbors and in close solidarity, were deemed separate: the arithmetical world, wherein one finds the so-called "spaces" without continuity, and the world of continuous magnitudes, "space" in the conventional meaning of the word. In this new vision these two worlds, formerly separate, comprise but a single unit. The embryonic vision of this Arithmetical Geometry (as I propose to designate the new geometry) is to be found in the Weil conjectures. In the development of some of my principal ideas, these conjectures were my primary source of inspiration, all through the years between 1958 and 1969.[11]

Weil's conjectures were very precise and very specific, and Weil could prove them in certain important special cases. But in full generality they were fantastically difficult because their very meaning could not be made precise without the development of a whole new theory that did not exist. One of Weil's conjectures proposed that the number of solutions of a diophantine equation—an equation whose solutions are required to be integers—could be found by a kind of algebra that was invented for the study of continuous functions. On the one hand, the desired information, the number of solutions of diophantine equations, pertained to the realm of the discrete—whole numbers. On the other hand, the claimed solution, "Betti numbers," are part of homology theory, which made sense only in the context of continuous manifolds (spheres, toruses, and their higher-dimensional analogs). Something very big was missing—a general theory that could bring together the discrete and the continuous, that could make the machinery of homology and cohomology, which was efficient and powerful in topology, valid in the remote realm of whole numbers.

The missing theory required generalizing the geometric notion of a space. Many spaces are considered in mathematics—Euclidean and non-Euclidean, the projective plane where a line at infinity is adjoined, curved manifolds both smooth and rough, Einstein-Minkowski space-time, knots and braids, pretzels and multipretzels, even the infinite-dimensional spaces used in quantum mechanics. They all can be thought of as sets of points, subject to conditions and constraints of one kind or another. But Grothendieck went further, to spaces without points. What sense does it make to talk about spaces without points?

The trick is to "algebrize" everything. Here is our own example (much simpler, of course, than what Grothendieck was concerned with). Out of pure algebra, one can create a familiar geometric space, the complex plane (the two-dimensional set of points represented by complex numbers of the form $a + bi$). Start with the set of all quadratic polynomials, $ax^2 + bx$

+ c, where a, b, and c are arbitrary real numbers. This set of polynomials is closed under addition and under multiplication by real numbers. Such an algebraic structure is called a "ring." Then we make the following agreement: if two of those polynomials differ by a multiple of the special polynomial $(x^2 + 1)$, we will regard them as "equivalent." In other words, we decide to treat $(x^2 + 1)$ as zero! The set of all scalar multiples of $(x^2 + 1)$ is called an "ideal," for some reason. By this equivalence relation, the ring of all polynomials in x is broken down into subsets. Each subset consists of polynomials equivalent to each other, meaning that the polynomials in each subset differ by a multiple of $(x^2 + 1)$. It then appears that there is a natural and convenient way to add and multiply these equivalence classes to one another. In fact, this algebraic structure of equivalence classes is isomorphic to the complex numbers!

Where does i, the square root of -1, come from? Well, a moment's thought will show that if $(x^2 + 1)$ is equivalent to 0, then x^2 is equivalent to -1, and so the equivalence class of $\{x\}$ works like i, the square root of -1. Thus, instead of defining complex numbers as points in the plane endowed with certain additive and multiplicative properties, we start with an algebra of polynomials and succeed at the end in constructing the space of complex numbers from the polynomial algebra. Thus a ring of polynomials, with a certain equivalence relation defined on it, turns out to be a space!

Here we are working with a very specific ring, the real polynomials in x. In more general contexts, including problems in number theory associated with diophantine equations, one again finds the algebraic structure, with operations of addition and multiplication, that is called a ring. Grothendieck's idea, roughly speaking, was to impose on *any ring at all* a superstructure constructed to have the axiomatic properties of what we usually call a space.

A standard, principal tool in algebraic geometry and topology was the "sheaf" invented by Jean Leray. A circle with a tangent line attached at every point is an elementary example of a sheaf. The next example is any smooth surface in 3-space with the variable tangent plane attached at each point. A third example is the same smooth surface, but with the line perpendicular to the tangent plane attached at each point. In fact, we can take any manifold (such as a higher-dimensional sphere or torus or multitorus) and at every point attach a plane, a hyperplane, or some other algebraic structure.

Grothendieck had the boldness to consider, with each geometric object such as a manifold, roughly speaking, the set of *all possible* attached algebraic objects, including lines or planes. He defined these huge structures axiomatically and called them "schemes" or "toposes." Then he showed how to do mathematics—prove theorems—about them. This was possible by using methods from a new, superabstract branch of mathematics called "category theory," which had recently been developed by Samuel Eilenberg, Saunders MacLane, and Henri Cartan, among others. In category theory, one doesn't start with *sets*, one starts with *mappings* ("arrows") connecting undefined objects. Grothendieck's toposes, then, are defined axiomatically to be like spaces with all possible sheaf structures. They don't have points. What they have is "cohomology"—a certain algebraic structure, that works to classify topological spaces.

For Grothendieck, a theorem had to be exactly right in every detail. Here is how he described his own way of working: "The first analogy that came to my mind is of immersing the nut in some softening liquid, and why not simply water? From time to time you rub so the liquid penetrates better, and otherwise you let time pass. The shell becomes more flexible through weeks and months—when the time is ripe, hand pressure is enough; the shell opens like a perfectly ripened avocado!"

Grothendieck wrote, "The two powerful ideas that had the most to contribute to the initiation and development of the new geometry are schemes and toposes. The very notion of a scheme has a childlike simplicity—so simple, so humble in fact that no one before me had the audacity to take it seriously."[12]

Grothendieck thinks of sheafs as

various "weights and measures" [which] have been devised to serve a general function, good or bad, of attaching "measures" (called "topological invariants") to those sprawled-out spaces which appear to resist, like fleeting mists, any sort of metrizability. . . . One of the oldest and most crucial of these invariants, introduced in the last century (by the Italian mathematician Betti) is formed from the various "groups" (or "spaces") called the "cohomology" associated with this space. . . . It was the Betti numbers that figure ("between the lines" naturally) in the Weil conjectures, which are their fundamental "reason for existence" and which give them meaning. Yet the possibility of associating these invariants with the "abstract" algebraic varieties that enter into these conjectures . . . that was something only to be hoped for.

He further wrote:

The new perspective and language introduced by the use of Leray's concepts of sheaves has led us to consider every kind of "space" and "variety" in a new light. The "new principle" that needed to be found was the idea of the topos. This idea encapsulates, in a single topological intuition both the traditional topological spaces, incarnation of the world of the continuous quantity, and the so-called "spaces" (or "varieties") of the unrepentant abstract algebraic geometers and a huge number of other sorts of structures which until that moment had appeared to belong irrevocably to the "arithmetical world" of discontinuous or "discrete" aggregates. Consider the set formed by all sheaves over a given topological space, or, if you like, the formidable arsenal of all the "rulers" that can be used in taking measurements of it. . . . We will treat the "ensemble" or "arsenal" as one equipped with a structure that may be considered "self-evident," one that crops up in front of one's nose, that is to say, a categorical structure. It functions as a kind of "superstructure of measurement," called the "category of sheaves" (over the given space). It turns out that one can "reconstitute" in all respects, the topological space by means of the associated "category of sheaves" (or arsenal of measuring instruments) . . . the idea of the topos had everything one could hope to cause a disturbance, primarily through its "self-evident" naturalness, through its simplicity . . . through that special quality which so often makes us cry out: "Oh, that's all there is to it" in a tone mixing betrayal with envy, that innuendo of the "extravagant," the "frivolous," that one reserves for all things that are unsettling by their unforeseen simplicity, causing us to recall, perhaps, the long buried days of our infancy.[13]

Grothendieck's Later Years

At this point we return to Grothendieck's chronological life history. In 1966 the International Congress of Mathematicians, meeting in Moscow, awarded him the Fields Medal, the highest international award for mathematical research. In protest against the repressive policies of the Soviet regime, he refused to attend to receive the prize.

In 1970, at the age of 42, still at the height of his creative powers and international fame, it is reported that Grothendieck learned that 5 percent of his institute's budget came from the French military. He demanded that the military money be refused. The administration of the IHES at first promised to do so but then accepted the money anyway. Grothendieck then left the institute, never to return.

The so-called "productive period" of my mathematical activity, which is to say the part that can be described by virtue of its properly vetted publications, covers the period from 1950 to 1979, that is to say 30 years. And, over a period of 25 years, between 1945 (when I was 17) and 1969 (approaching my 42nd year) I devoted virtually all of my energy to research in mathematics. An exorbitant investment, I would agree. It was paid for through a long period of spiritual stagnation, by what one may call a burdensome oppression which I evoke more than once in the pages of *Récoltes et Semailles* . . . the greater part of my energy was consecrated to what one might call detail work: the scrupulous work of shaping, assembling, getting things to work, all that was essential for the construction of all the rooms of the houses, which some interior voice (a demon perhaps?) exhorted me to build. . . .

In evaluating his contribution to the mathematics of the 20th century, Grothendieck describes the feverish pace at which he worked and his belief, shared by many, that he had created an extraordinary new structure of ideas, particularly in algebraic geometry:

I rarely had the time to write down in black and white, save in sketching the barest outlines, the invisible master-plan that except (as it became abundantly clear later) to myself underlined everything, and which, over the course of days, months and years guided my hand with the certainty of a somnambulist. . . And by hundreds, if not thousands of original concepts which have become part of the common patrimony of mathematics, even to the very names which I gave them when they were propounded. . . . In the history of mathematics I believe myself to be the person who has introduced the greatest number of new ideas into our science. . . . The theme of schemas, their prolongations and their ramifications, that I'd completed at the time of my departure, represents all by itself the greatest work on the foundations of mathematics ever done in the whole history of mathematics and undoubtedly one of the greatest achievements in the whole history of Science.[14]

In 1975 Weil's last conjecture was finally proved by Grothendieck's student Pierre Deligne, after Grothendieck had departed from the mathematical world. His proof was a continuation of Grothendieck's work but relied on a result from classical mathematics that Grothendieck would not have known about. Grothendieck expressed bitter disapproval. Deligne's method for finishing the proof did not follow Grothendieck's grander and more difficult plan.

For a few years Grothendieck devoted himself ardently to the cause of the environment. With the cooperation of fellow mathematicians Pierre Samuel and Claude Chevalley, he founded an ecology-saving organization, *Vivre et Survivre*. The Bourbaki member Pierre Cartier shared Grothendieck's environmental concerns but complained that Grothendieck talked politics whenever he was invited to lecture on mathematics. This irritated everyone who came to hear him—even if they agreed with his politics. At the International Congress of Mathematicians in Nice in 1970, Grothendieck handed out leaflets and tried to set up a table for *Vivre et Survivre*. His old friend and collaborator, Jean Dieudonné, now a Dean at the University of Nice and one of the organizers of the Congress, prohibited Grothendieck's table. So Grothendieck put up a table on the street outside the meeting. The chief of police showed up and asked Grothendieck to move his table back a few yards off the sidewalk. But he refused. "He wanted to be put in jail," Cartier recalled. "He really wanted to be put in jail!" Finally the table was moved back far enough to satisfy the police.

During that time he had a position at the Collège de France in Paris, and then he had one at the University at Montpellier, where he had once been a student. David Ruelle, a leading mathematical physicist who was Grothendieck's colleague at the IHES starting in 1964, has recently written that "Grothendieck's program was of daunting generality, magnitude, and difficulty. In hindsight we know how successful the enterprise has been, but it is humbling to think of the intellectual courage and force needed to get the project started and moving. We know that some of the greatest mathematical achievements of the late twentieth century are based on Grothendieck's vision. . . .

Our great loss is that we don't know what other new avenues of knowledge he might have opened if he had not abandoned mathematics, or been abandoned by it."[15]

He continues, "It may be hard to believe that a mathematician of Grothendieck's caliber could not find an adequate academic position in France after he left the IHES. I am convinced that if Grothendieck had been a former student of the École Normale and if he had been part of the system, a position commensurate with his mathematical achievements would have been found for him. . . . Something shameful has taken place. And the disposal of Grothendieck will remain a disgrace in the history of twentieth-century mathematics."[16]

In May 1972, while visiting Rutgers University in New Jersey, Grothendieck met a young graduate student in mathematics named Justine (now named Justine Bumby). When he returned to France, she went with him. They lived together for 2 years and had a son John, who is now a mathematician. Once, while she was with Grothendieck at a peaceful demonstration in Avignon, the police came and harassed the demonstrators. Grothendieck got angry when they started pestering him. Justine recalled in an interview with Allyn Jackson: "The next thing we know, the two policemen are on the ground." Grothendieck, who once practiced boxing, had single-handedly decked two police officers. At the police station, the chief of police expressed his desire to avoid trouble between police and professors, and Grothendieck and Justine were released.

For a long time Grothendieck was hospitable to all sorts of marginal "hippies." In 1977 he was indicted and tried under a 1945 regulation that made it a misdemeanor to meet with a foreigner. He was given a suspended sentence of 6 months in prison and a fine of 20,000 francs (about $40).

He continued writing about mathematics, and circulating ideas among his acquaintances. The 1000-page *Récoltes et Semailles* was composed between 1983 and 1988. It's a remarkable document, full of beautiful images and inspired rhetoric. Much of it is devoted to attacks on his pupils and followers, both for failing to carry out the projects he left for them and for shamelessly utilizing morsels, bits, and pieces of his work for their own ends.

In 1988 Sweden awarded him the $160,000 Crafoord Prize, to be shared with Pierre Deligne. Grothendieck refused the prize and wrote:

> The work that brought me to the kind attention of the Academy was done twenty-five years ago at a time when I was part of the scientific community and essentially shared its spirit and its values. I left that environment in 1970, and, while keeping my passion for scientific research, inwardly I have retreated more and more from the scientific milieu. Meanwhile, the ethics of the scientific community (at least among mathematicians) have declined to the point that outright theft among colleagues (especially at the expense of those who are in no position to defend themselves) has nearly become the general rule, and is in any case tolerated by all, even in the most obvious and iniquitous cases. . . . I do not doubt that before the end of the century totally unforeseen events will completely change our notions about science and its goals and the spirit in which scientific work is done. No doubt the Royal Academy will then be among the institutions and the people who will have an important role to play in this unprecedented renovation, after an equally unprecedented civilization collapse.[17]

People who had been close to him were disturbed by this cryptic announcement of an apocalypse, even more than by his harsh condemnation of the mathematical world. Then, in 1992, he disappeared, cutting off all known connections with family and friends.

Grothendieck's resentment toward his colleagues and former students is surprising in a man who wrote lovingly in earlier times about his collaborative work. The fact that many individuals worked hard to write up his discoveries also speaks of a personality that was appreciated and admired by his peers.

But the course of his later life, including his growing dissatisfaction with an existence limited to research in mathematics, invites reflection. One way to think about some of his paranoid accusations, and his increasing withdrawal from his family and friends, is to speculate that his personal pain and struggles had been kept under control while he was devoting every waking hour to mathematics. That was his joy and his solace, as well as an escape from the tragic events of his childhood. While engaged in mathematics, he participated in a world that had order and beauty and was more predictable than the uprooting and losses during his early years. It was a world in which he could share his thoughts and his discoveries and put to work his incredible energy and creativity.

But however closely he immersed himself in the mathematical structures he created, it is likely that even during his most productive years he was haunted by the war years. Thinking of the many reasons he could have had to reject ordinary life, one can understand his seeking uninterrupted absorption in mathematics. But that total absorption would have excluded a more sustainable balance between his diverse interests, loves, and connection to his work. It is possible that witnessing the police brutalities against the Algerians, together with his fear of an environmental disaster, resulted in a fracture from the bounded abstract world in which he had lived productively for 20 years. His yearning for such a balance runs through many passages of *Récoltes et Semailles:*

> This work of discovery, the concentrated attention involved and its ardent solicitude, constituted a primeval force, analogous to the sun's heat and the germination and gestation of seeds sown in the nourishing earth and for their miraculous bursting forth into the light of day. . . . I've made use of the images of the *builder* and of the *pioneer* or explorer . . . in this "male builder's" drive which, would seem to push me relentlessly to engineer new constructions, I have, at the same time, discerned in me something of the homebody, someone with a profound attachment to "the

home." Above all else, it is "his" home, that of persons "closest" to him—the site of an intimate living entity of which he feels himself a part. . . . And, in this drive to "make" houses (as one "makes" love . . .) there is above all, tenderness. There is furthermore the urge for *contact* with those materials that one shapes a bit at a time, with loving care, and which one only knows through that loving contact. . . . Because the *home*, above all and secretly in all of us, is the *Mother*—that which surrounds and shelters us, source at once of refuge, and comfort. . . .[18]

In September 1995 it was announced (by Roy Lisker) that Jean Malgoire of the University of Montpellier had visited Grothendieck, in a village in the Pyrenees Mountains. There Grothendieck meditates and sustains himself in a manner totally harmless to the environment. In Paris there is a Grothendieck circle, collecting, translating, and publishing his writings, while Grothendieck maintains his total distance from his previous mathematical life. Recently he demanded that they stop all work on his publications.

Into Madness

Mathematics involves working just with pencil and paper, or perhaps with chalk or a computer keyboard. And yet it has been wisely said that mathematics can be a very dangerous profession—dangerous especially to some of its more vulnerable practitioners. As a graduate student, I (R.H.) heard a lecture on A-contractive mappings by a Rutgers professor named Wolodymyr V. Petryshyn. Professor Petryshyn was later incarcerated. He actually murdered with a hammer his beloved wife Arcadia Olenska-Petryshyn, who was an artist of some reputation.[19] According to those who knew him, Petryshyn had discovered a mistake in his estimable book *Generalized Topological Degree and Semilinear Equations*. His publishers assured him that it was not such a serious problem. He, however, felt that his omission of a necessary hypothesis meant total disgrace. Felix Browder, who had known Petryshyn for decades, was quoted: "It sounds as if his perfectionism drove him to insanity."

When I (R.H.) arrived at Stanford in 1962 for the start of a 2-year instructorship, I met Professor Karel de Leeuw. Karel sat watching as I unpacked my books. He was well known in the department for his exceptional interest in students as individual human beings. Every year he ended his graduate course with a party at his house. But one of the Stanford math graduate students, Ted Streleski, decided that in retribution for his 19 years of suffering there without progressing to a degree, it was "completely logical" to murder Karel de Leeuw. He explained to a newspaper, "Stanford University took 19 years of my life with impunity, and I decided I would not let that pass."

Maybe it's morbid to recount such sad stories. We are telling them because total immersion in mathematical life, with the expectation of perfection, can contribute to madness. Petryshyn's eruption was mad, but to a fellow mathematician it is in some degree understandable. We all know that our results, our publications, are supposed to be completely correct, logically irrefutable. We also know that in very many cases, even though everything looks right, there remains an aching uncertainty. Is it really absolutely correct? Haven't I overlooked something? What if it turns out to be all wrong? With this persisting anxiety goes an underlying feeling that if that should actually be the case, then disaster and disgrace will be utter and final. All this is in fact delusion, yet it is somehow part of the ethos that we absorb somewhere along the line in our training and indoctrination. "It's either right or wrong. You're supposed to know the difference."

So a particularly rigid, inflexible personality could experience a sense of total destruction if his support in perfectly rigorous mathematics turned out to be cracked and broken. And then— what then? What then?

Although mathematicians increasingly work in pairs and in groups, traditionally we mostly worked alone. To some, working alone is a matter of pride, of personal identity. But working alone makes you especially vulnerable. A partner may catch your mistake. On your own, you can kid yourself and go on and on in error. If you believe you know what is right, what is a proof, you may follow yourself into the flames. Something that is less likely to happen in a kind of work that is inherently cooperative and social.

Ted Kaczynski

In 1996 and 1997, the most famous mathematician in America was Ted Kaczynski. A Ph.D. from the University of Michigan and an ex–faculty member at Berkeley, he had not published or taught in the preceding years. He certainly was not welcomed by the mathematics profession as a representative. But he was trained and qualified as a mathematician, and his particular kind of insanity would not be found in a salesperson or an auto mechanic.

For those who don't remember the story, the Unabomber was a mystery man for years. From 1978 to 1995, someone was mailing bombs to American scientists and doing serious damage. David Gelernter at Yale was permanently disfigured when he opened one of these nasty surprise packages. Gilbert Murray, president of the California Forestry Association, was killed. In 1995 the secret Unabomber demanded that his Manifesto be published in the *New York Times*.

We try to summarize his reasoning, which could even be called a theorem. We present this, not only for its intrinsic interest but also to show the mathematical nature of his obsession.

1. Industrialization is destroying the natural world, which gave rise to the human race, and on which the human race depends for its existence. Already very severe, even

Figure 4-2. Ted Kaczynski, before he became the Unabomber. Courtesy of Archives of the Mathematisches Forschungsinstitut Oberwolfach.

irreparable damage has been done to the natural world: extinction of species, depletion of resources, universal pollution by many different pollutants.

2. Increasing population and the demand of capital for constant growth generate more destruction of the natural world, at an ever accelerating rate. We can see catastrophe within decades.

3. This oncoming catastrophe far outweighs any other considerations of humanity, morality, or tradition. It is the duty of every thinking person to understand it, and to act on it, immediately, and as effectively as possible.

4. But the short term self-interests of government, communications media, political parties, induce them all to close their eyes, and turn away from this overwhelmingly threatening reality.

5. Any measures to wake people up, to force them to pay attention, are justified, even demanded.

6. Threatening to kill or actually killing people who are implicated in this dreadful process is one way, maybe the only way, to get the needed attention to this cause.

7. Since I, Ted Kaczynski, do understand this, it is my absolute duty to proceed with threats and actual murders for this all-important end, to save humanity and the earth.

Reading the manifesto, the mathematician John Allen Paulos guessed from the tone, content, and structure that its author was a mathematician. He wrote an op-ed piece saying so for the *New York Times*. That angered some mathematicians who thought it was bad for their image, and the *Wall Street Journal* published a long article on the fracas. In his essay Paulos had opined that Kaczynski's Ph.D. in mathematics was "perhaps not quite as anomalous as it seemed (despite the fact that mathematicians are for the most part humorous sorts, not asocial loners, and that the only time most of us use the phrase 'blow up' is when we consider division by zero)".

Doron Zeilberger, a mathematician at Rutgers University, was prompted to look up Ted Kaczynski's papers. He reported: "They are paragons of precision! They are written in a no-nonsense terse, yet complete step-by-step style, that makes most math papers look like long-winded sociology. The math is also beautiful! What a shame that he quit math for bombing." Note that while it's already a shame that Ted Kaczynski became a bomber, it will become a still greater shame when we realize that his publications showed real promise!

Kaczynski's brother, David, recognized the mind behind the manifesto and the bombs and did the painful but necessary task of turning his brother in to the Federal Bureau of Investigation. Ted Kaczynski was found in his hermit cabin in Montana, where for years he had been thinking hard how to prevent humanity from destroying the planet. He is quoted in the *New York*

Times, January 23, 1998: "My occupation is an open question. I was once an assistant professor of mathematics. Since then, I have spent time living in the woods of Montana."

He is now serving a life sentence in a high-security federal prison. There he continues to correspond with like-minded anti-industrialization fanatics.

Since the conclusion of his manifesto is murder, surely it can be carried out only if there is absolute certainty that it is correct. Such absolute certainty is the special characteristic of mathematical reasoning. If Kaczynksi had not been a mathematician, he might still have believed in an imminent environmental disaster. He might still have committed murder, following his beliefs. But surely his case would not have been argued so rationally.

The answer to his argument, I believe, is simply the age-old cry of Oliver Cromwell: "I beseech you in the bowels of Christ to consider that you may be mistaken!"

André Bloch

The French mathematician André Bloch lived from 1893 to 1948 and won the Becquerel Prize for discovering Bloch's constant which is important in the theory of univalent analytic functions of a complex variable. He was a patient of Henri Baruk, who was, according to Cartan and Ferrand, "one of the greatest French psychiatrists of the mid-twentieth century." In Baruk's autobiography, *Patients Are People Like Us*, he writes of Bloch:

> Every day for forty years this man sat at a table in a little corridor leading to the room he occupied, never budging from his position, except to take his meals, until evening. He passed his time scribbling algebraic or mathematical signs on bits of paper, or else plunged into reading and annotating books on mathematics whose intellectual level was that of the great specialists in the field. . . . At six-thirty he would close his notebooks and

books, dine, then immediately return to his room, fall on his bed and sleep through until the next morning. While other patients constantly requested that they be given their freedom, he was perfectly happy to study his equations and keep his correspondence up to date.[20]

Unlike other patients, Bloch refused to go out of the building onto the grounds, saying "Mathematics is enough for me." Bloch completed a substantial body of work: four papers on holomorphic and meromorophic functions and short articles on function theory, number theory, geometry, algebraic equations, and kinematics. He was self-taught, for his studies had been brutally and prematurely interrupted by World War I. His last paper was a collaboration with another mathematician who had been hospitalized for a short while with him in the Maison de Charenton.

A few months after Bloch's death in 1948, his story was told by Professor Georges Valiron at the annual meeting of the Societé pour l'Avancement des Sciences. "In 1910 I had both (André and his brother Georges) in my class (at the École Polytechnique.) They left in 1914 because of the war." Professor Valiron's listeners needed no reminders of that war—its trench warfare and suicide attacks, its misery, degradation, and horror. He went directly to the outcome. "On November 17, 1917, at the end of a leave and three days before his 24th birthday, in the course of a crisis of madness he killed his brother Georges, his uncle, and his aunt. Declared insane, he was confined to the mental hospital at Saint Maurice, where he remained until his death on October 11, 1948."

After months at the front, André Bloch had fallen off an observation post during a bombardment, and the shock made him unfit for active service. His brother Georges, whom he killed, had been wounded in the head and lost an eye. The murders took place at a meal in the family apartment on the Boulevard de Courcelles in Paris. Afterward André ran screaming into the street and let himself be arrested.

When Dr. Baruk read his patient's history, he found it difficult to imagine that such a charming, cultivated, polite man could have committed such an act. "He was the model inmate, whom everyone in the hospital loved."[21] One day a younger brother of André appeared at the hospital. He had been living in Mexico, was passing through Paris, and wanted to see André. André gave no sign of affection or welcome to his brother. His manner was extremely cold. . . . The next day he explained to Dr. Baruk: "It's a matter of mathematical logic. There were mentally ill people in my family, on the maternal side, to be exact. The destruction of the whole branch had to follow as a matter of course. I started my job at the time of the famous meal, but never got a chance to finish." Dr. Baruk told Bloch that his ideas were horrifying. "You are using emotional language," Bloch answered. "Above all there is mathematics and its laws. You know very well that my philosophy is based on pragmatism and absolute rationality." Dr. Baruk diagnosed "morbid rationalism . . . a crime of logic, performed in the name of absolute rationalism, as dangerous as any spontaneous passion."[22]

André Bloch was an extreme example. He worked on mathematics problems for decades, for the same hours every day, sitting in the same corner in a corridor in the Charenton asylum. He was able to do important mathematical work. But his social and ethical judgment was totally impaired, akin to that of a patient suffering from prefrontal cortical injury. His was an extreme and rare case. Nevertheless, it is a useful reminder of the harm that can come from isolating abstract knowledge from real-life uses, or from constructing a life totally devoted to strenuous mental efforts without social connections or diverse interests.

Kurt Gödel

The greatest logician of the 20th, or any other century, also could have been diagnosed with morbid rationalism, although he certainly never killed any one or even raised his hand in anger.

When Gödel went to a U.S. judge to attain his citizenship, accompanied by his friend Oskar Morgenstern, he insisted on correcting the judge when the judge said that events like those in Germany, resulting in Hitler's becoming dictator, could never happen under the U.S. constitution. Gödel had noticed a way that this could happen under the constitution! Morgenstern managed to deflect his incipient lecture, so that the citizenship proceedings could continue. In his last years, after Morgenstern and his other friend at the Institute for Advanced Study, Albert Einstein, had died, Gödel simply didn't talk to anybody—that is, talk about his real interests and concerns, mathematics and philosophy. He wrote papers for his desk drawer. His wife, Adele, of course prepared his meals. Unfortunately, she became ill and had to go to a hospital. Without her protection, he had no proof that any food he was offered was safe. He cleverly prevented himself from being poisoned by refusing to eat anything at all. Adele returned home from the hospital at the end of December and persuaded Gödel to enter Princeton Hospital. Kurt Gödel died at age 72, in the fetal position, at 1:00 in the afternoon, on Saturday, January 14, 1978. He weighed 65 pounds. According to the death certificate, on file in the Mercer County Courthouse in Trenton, he died of malnutrition and inanition caused by personality disturbance.

Thus perished the greatest logician of all time! And what about Adele? How long did she survive? How did she endure decades of serving her genius husband in total isolation from other human contact? What did she do? What did she think?

In looking back over these disturbing stories, we can't help but remember Streleski's and Bloch's explanations: their crimes were completely logical! Indeed, greater crimes are committed, by politicians and generals, also claiming that their actions are "completely logical."

It is really claimed by some philosophers that the propositional and predicate calculi—modern formal logic—are infallible (e.g., John Worrall and Elie Zahar in editing Lakatos' *Proofs*

Figure 4-3. Kurt Gödel and Adele Gödel. Courtesy of the Kurt Gödel Papers, The Shelby White and Leon Levy Archives Center, Institute for Advanced Study, Princeton, NJ, USA, on deposit at Princeton University.

and Refutations). "From true premises, true conclusions follow, infallibly." How dangerous this dogma can be! Logic can never be anything but a tool, an action, or a procedure carried out by a *human being* (or by a machine created and programmed by a human being). Logic, such an essential tool of science and philosophy, sometimes becomes a sort of false god, outranking the most fundamental human impulses, such as "Thou shalt not kill." Or even, "Eat to stay alive."

In answer to our question, "Can mathematics become a dangerous addiction?" we must answer, "Yes, to a susceptible mind, under unfavorable conditions, that has indeed happened."

Bibliography

Aczel, A. D. (2006). *The artist and the mathematician*. New York: Avalon Publishing Group.

Baruk, H. (1978). *Patients are people like us*. New York: William Morrow.

Campbell, D. M. (1985). Beauty and the beast: The strange case of André Bloch. *Mathematical Intelligencer* 7(4), 36–38.

Cartan, H., & Ferrand, J. (1988). The case of André Bloch. *Mathematical Intelligencer, 10*(1), 241.

Cartier, P. (2001). A mad day's work: From Grothendieck to Connes and Kontsevich. *Bulletin of the American Mathematical Society 38*(4), 389–408.

Goldstein, R. (2006). *Incompleteness*. New York: W. W. Norton.

Grothendieck, A., Colmez, P. (Ed.), & Serre, J.-P. (2001). *Grothendieck-Serre correspondence*. Paris: Societé Mathematique Francaise. Bilingual edition, Providence, R.I.: American Mathematical Society. 2004.

Grothendieck, A. (1986). *Récoltes et Semailles*. Unpublished manuscript.

Grothendieck, A. (1989). Letter refusing the Crafoord Prize, *Le Monde*, May 4, 1988. *Mathematical Intelligencer* 11(1), 34–35.

Jackson, A. (2004). Grothendieck. *Notices of the American Mathematical Society* (Oct./Nov.), 1038–1056, 1196–1212.

Jackson, A. (1999). The IHES at forty. *Notices of the American Mathematical Society* 46(3), 330.

Lisker, R. (Translator) (1990). *Ferment* vol. V(5), June 25. The quest for Alexandre Grothendieck; #6 October 1: Grothendieck, 2; #7 October 25: Grothendieck 3; #8 November 27; Grothendieck 4; #9 January 1: Grothendieck 5. These are also contained in a book entitled *The quest for Alexandre Grothendieck*, available from the author. Translation of the first 100 pages of *Récoltes et Semailles*. Conditions for obtaining these may be read at http://www.fermentmagazine.org/home5.html Grothendieck circle website http://www.grothendieckcircle.org July 20, 2007.

Nasar, S. (1998). *A beautiful mind.* New York: Touchstone.

Paulos, J. A. (1998). *Once upon a number.* New York: Basic Books.

Ruelle, D. (2007). *The mathematician's brain.* Princeton, N.J.: Princeton University Press.

Schneps, L. (2008). *Grothendieck-Serre correspondence,* book review. *Mathematical Intelligencer 30*(1), 60–68.

Schwartz, L. (2001). *A mathematician grappling with his century.* Basel: Birkhäuser.

Weil, A. (1992). *The apprenticeship of a mathematician.* Basel: Birkhäuser.

Zeilberger, D. Blog, Internet.

+ 5 +

Friendships and Partnerships

Do mathematicians have friends? The popular image of a mathematician is that of a solitary man, alone at his desk or blackboard. In this chapter, we will see how far from the truth this picture is. While the sustained concentration needed for mathematical research does require quiet and a highly focused mindset, an intense and prolonged independent search can come to a dead end. The researcher who listens to colleagues' insights can break through his private perspective.

One morning early in my (Hersh's) years as a thesis student of Peter Lax, I entered my mentor's office to find him glowing in smiles. "Louis is back!" he cried out to me. Louis, I wondered? Oh yes, Louis Nirenberg, also one of the partial differential specialists on the faculty of NYU's Courant Institute. He had been on leave in England; now he was back home! At the time. I didn't get it. Louis Nirenberg and Peter Lax were grad students together at NYU. Then they both stayed on to become famous faculty members there—Louis, a world master at elliptic partial differential equations (PDEs), and Peter, a world master at hyperbolic PDEs. They hardly ever collaborated or produced joint publications. But their conversations and their intellectual and emotional interactions were a vital part of their creativity and success.

In the history of mathematics many names occur in pairs—Hardy and Littlewood, Cayley and Sylvester, Weierstrass and Kovalevskaya, Polya and Szegö, Riesz and Nagy, Hardy and Ramanujan, Minkowski and Hilbert, and Lax and Phillips. Each of these pairs was different. Karl Weierstrass and Sonia Kovalevskaya were not merely mentor and pupil, they were deeply involved with each other emotionally. David Hilbert and Hermann Minkowski were close friends who gave each other vital support and stimulation, although they never collaborated on a publication. G. H. Hardy and J. E. Littlewood coauthored nearly 100 papers, over 35 years, yet they seldom met face to face. Littlewood said, "Hardy was unhappy except with bright conversation available. . . . Our habits were about as opposite as could be."[1]

We take a historical route to describe some of the most interesting friendships among colleagues and across generations. Then we tell about two very different mathematical marriages— between Grace Chisholm and Will Young, and between Julia Bowman and Raphael Robinson. We conclude our survey of mathematical friendships with a section on friendships among women, including Olga Taussky-Todd and Emmy Noether.

Mentors

Two mathematicians can relate as teacher and student, as collaborators, or as friends. A vital relationship in the lives of most successful mathematicians has been as an apprentice to a mentor. One of the most famous apprentice-mentor stories is about a 20-year-old married Russian beginner, Sonia Kovalevskaya, and a 55-year-old master of complex analysis, the bachelor Karl Weierstrass. Kovalevskaya arrived in Berlin in October 1870. She obtained an audience with the great Professor Weierstrass and pleaded with him to accept her as a student. Such a thing was then impossible. Nevertheless, Weierstrass gave her a few test problems. When she returned a week later, her solutions were not only correct, but they were also original. Weierstrass said that her paper showed "the gift of intuitive genius to a degree he had seldom found among even his older and more developed students."[2] So Weierstrass recruited the prominent physiologist Emil DuBois-Reymond, the famous pathologist Rudolf Virchow, and the renowned physicist-physiologist Hermannn Helmholtz to join him in requesting permission for Sonia to register as a student at the University of Berlin. But still the faculty senate voted no! So Weierstrass offered instead to recapitulate his lectures for her and moreover to tell her about his own research. For 4 years she visited him every Sunday that she was in Berlin. And he came once a week to the apartment that she shared with a friend.[3] Kovalevskaya said, "These studies had the deepest possible influence on my mathematical career. They determined finally and irrevocably the direction I was to follow in my later scientific work; all my work has been done precisely in the spirit of Weierstrass."[4]

Weierstrass's role in Kovalevskaya's scientific and personal affairs far transcended the usual teacher-student relationship. He found her a refreshingly enthusiastic participant in all his thoughts, and ideas that he had fumbled for became clear in his conversations with her. In 1873, while he was on vacation in Italy, Weierstrass wrote to Sonia: "During my stay here I have thought about you very often and imagined how it would be if only I could spend a few weeks with you, my dearest friend, in such a magnificent natural setting. How wonderful it would be for us here—you with your imaginative mind, I—stimulated and refreshed by your enthusiasm—dreams and flights of fancy about so many puzzles that remain for us to solve about finite and infinite spaces, the stability of the solar system, and all the other great problems of the mathematical physics of the future. However, I learned long ago to resign myself if not every beautiful dream becomes true." It seemed to him that they had been close throughout his entire life and "never have I found anyone who could bring me such understanding of the highest aims of science, such joyful accord with my intentions and basic principles as you."[5]

Yet their relationship did not remain untroubled. Weierstrass disapproved of Sonia's links with socialist circles, her literary

Figure 5-1. Karl Weierstrass, the great German analyst.

work, and her advocacy of the emancipation of women. Many of his letters to her went unanswered; at one juncture she did not respond for 3 years. . . . Sonia Kovalevskaya died on February 10, 1891, in her fortieth year, of a pulmonary infection. Her untimely death, when she was still in her prime, caused Weierstrass much grief. He burned all her letters to him.[6]

Another famous friendship was between the great German mathematician David Hilbert and his colleague Hermann Minkowski. Minkowski, born in 1864, was the son of a Russian merchant. He and Hilbert first met in gymnasium, and then became close friends as students at the University of Königsberg.

Although the two had quite different personalities, they shared an enthusiastic love of mathematics and a deep, fundamental optimism.[7] Hilbert openly expressed his need for Minkowski's companionship and stimulation. Together with their young teacher, Adolf Hurwitz, they met at 5:00 every afternoon. They formed a friendship for life.[8] Realizing the many benefits

Figure 5-2. David Hilbert. Source: *Mathematical People:
Profiles and Interviews*. Eds. Donald J. Albers and
G. L. Alexanderson. Boston: Birkhauser, 1985. Pg. 285.
Reprinted by kind permission of Springer Science and
Business Media.

of their relationship, these young mathematicians tried hard
to stay geographically close. Even when separated, they relied
heavily on each other's advice. In 1900 Hilbert was preparing
his famous speech to the International Congress of Mathematicians in Paris, where he planned to outline 23 problems for
mathematicians to solve in the coming century. Even though
Minkowski was in Zurich, he was eager to help. The two of
them, and Adolf Hurwitz, corresponded frequently on the form
and content of the lecture. Hilbert paid close attention to their

criticisms of his address, "Mathematical Problems," until he reached the final draft. At the turn of the century, the University of Göttingen became the leading center of mathematical work under Hilbert's leadership. But not until he was able to arrange a professorship for Minkowski was Hilbert truly satisfied. After Minkowski arrived in Göttingen in the fall of 1902, Hilbert was no longer lonely. He needed only a telephone call, or a pebble tossed up against the little corner window of his study, and he was ready to join his friend.[9]

The relationship between these two men dazzled and instructed those around them. Max Born recalled, "The conversation of the two friends was an intellectual fireworks display. Full of wit and humor and still of deep seriousness."[10] Minkowski provided a critical ear as Hilbert prepared his classes; they taught a joint seminar; they shared their interest in physics. The friendship came to a tragic end when Minkowski, at the age of 44, died of a ruptured appendix. We have written more about Göttingen in the chapter on mathematical communities.

Hardy, Littlewood, and Ramanujan

G. H. Hardy was the most influential English mathematician of the first quarter of the 20th century. (His lectures won Norbert Wiener from philosophy to mathematics.) Hardy's beautifully written essay *A Mathematician's Apology* is widely known. His apology is for a life devoted to pure mathematics. His defense is that, although he never did anything "useful," he did add to the world's knowledge, guided by beauty as well as truth. He wrote, "The mathematician's patterns, like the painter's or the poet's, must be beautiful; the ideas, like the colours or the words, must fit together in a harmonious way. Beauty is the first test: there is no permanent place in the world for ugly mathematics."[11] (In chapter 2, we took up the question of beauty in mathematics.)

The essay is melancholy. At 60, says Hardy, he is too old to have new ideas. That is why he is reduced to writing books instead of doing a mathematician's proper business, discovering and creating new mathematics. In a foreword to a posthumous reprinting of Hardy's essay, C. P. Snow painted a moving picture of Hardy's life. (A physicist turned scientific bureaucrat, Snow wrote novels about political maneuvering among academics, and a much-quoted lecture and subsequent book, *The Two Cultures*, bemoaning the literati's ignorance of physics.)

In his prime, wrote Snow, Hardy lived among some of the best intellectual company in the world. He was one of the most outstanding young men in his circle. "His life remained the life of a brilliant young man until he was old; so did his spirit: his games, his interests, kept the lightness of a young don's. And like many men who keep a young man's interests into their sixties, his last years were the darker for it."[12]

Hardy found English analysis and number theory a quiet backwater and raised them to a Continental standard. He wrote: "I shall never forget the astonishment with which I read Jordan's famous *Cours d'analyse*, the first inspiration for so many mathematicians of my generation, and learnt for the first time as I read it what mathematics really meant. The real crises of my career came ten or twelve years later, in 1911, when I began my long collaboration with Littlewood, and in 1913, when I discovered Ramanujan. All my better work since then has been bound up with theirs, and it is obvious that my association with them was the decisive event of my life."[13]

Hardy's collaboration with Littlewood began in 1911 and lasted 35 years. All of Hardy's major work was done with Littlewood or Ramanujan. The Hardy-Littlewood partnership dominated English pure mathematics for a generation. Hardy said Littlewood "was the man most likely to storm and smash a really deep and formidable problem; there was no one else who could command such a combination of insight, technique and power."[14] The number theorist Edmund Landau said, "The mathematician Hardy-Littlewood was the best in the world, with Littlewood the more original genius and Hardy the better journalist."[15]

Figure 5-3. Three influential mathematicians in conversation: Richard Courant, G. H. Hardy, and Oswald Veblen. Courtesy of Archives of the Mathematisches Forschungsinstitut Oberwolfach.

Littlewood is now remembered best for his prodigious productivity in his later years. But in his prime he was the antithesis of the stereotypical mathematician: muscular and athletic, a proficient rock climber. He was a bachelor, like Hardy, but unlike Hardy, he appreciated female companionship. It was a well-known "secret" that a young woman he called his niece was actually his daughter, by a liaison with a colleague's wife.

There is a fascinating contrast between the emotional tone of Hardy's two great friendships—with his English collaborator John Littlewood and with his Indian collaborator Srinivasa Ramanujan.

How different was Hardy's way with Littlewood, from Hilbert's way with Minkowski! The Danish mathematician Harald Bohr (brother of the physicist Niels) reported:

When Hardy once stayed with me in Copenhagen, thick mathematical letters arrived daily from Littlewood, who was obviously very much in the mood for work, and I have seen Hardy calmly throw the letters into a corner of the room, saying, "I suppose I shall want to read them some day." This was according to one of the "axioms" of the Hardy-Littlewood collaboration: "When one received a letter from the other, he was under no obligation whatsoever to read it, let alone to answer it—because, as they said, it might be that the recipient of the letter would prefer not to work at that particular time, or perhaps that he was just then interested in other problems."[16]

Hardy's student Mary Cartwright observed the Hardy-Littlewood relationship from another angle. When Hardy returned to Cambridge from Oxford as the Sadleirian chair, she asked him if he would be offering a seminar similar to the Friday evening sessions she had enjoyed at Oxford. He replied that he would probably come to some arrangement with Littlewood. Soon after, the lecture list announced a Hardy-Littlewood class, to meet in Littlewood's rooms.[17]

The Oxford mathematician E. C. Titchmarsh (1899–1963) thought this seminar was a model of what such a thing should be. Mathematicians of all nationalities and ages were encouraged to present their own work, and the whole seminar was delightfully informal, with free discussion after each paper.[18] Cartwright recalled that at the first Hardy-Littlewood class, Littlewood was speaking. Hardy came in late, helped himself liberally to tea and began to ask questions, as if he were trying to pin Littlewood on details. Littlewood told Hardy that he was not prepared to be heckled. Thenceforth, Hardy and Littlewood alternated classes. Cartwright does not recall them being present together at any subsequent class. Eventually, Littlewood stopped participating, though the class continued to meet in his rooms. The class became known as "the Hardy-Littlewood Conversation Class at which Littlewood is never present."[19] Hardy and Littlewood, while deeply connected mathematically, had little

Figure 5-4. Dame Mary Cartwright, the famous British analyst. Courtesy of Archives of the Mathematisches Forschungsinstitut Oberwolfach.

apparent affection for each other. Their only interaction was by mail around their deep, difficult mathematics!

On one famous occasion, they did have to meet. They had to deal with an unprecedented human problem. Hardy had received a letter from an unknown person far away in India:

> Dear Sir, I beg to introduce myself to you as a clerk in the Accounts Department of the Port Trust Office at Madras on a salary of only 20 pounds per annum. I am now about 23 years of age. I have had no University education, but I have undergone the ordinary school course. After leaving school I have been employing the spare time at my disposal to work at Mathematics. I have not trodden through the conventional regular course which is followed in a University course, but I am striking out a new path for myself. I have made a special investigation of divergent series in general and the results I get are termed by the local mathematicians as "startling."[20]

The occasion demanded an actual conversation with Littlewood, who lived nearby in other rooms at Cambridge. Together Hardy and Littlewood concluded that Ramanujan was not an impostor but a genius mathematician. (Letters from Ramanujan had been ignored by two other leading English mathematicians.)

Hardy quickly worked to bring Ramanujan to England. This was a delicate matter because Ramanujan was an observant Hindu and his religion forbade him to cross the ocean. At last, however, he accepted Hardy's invitation. He spent 3 years in England during World War I, working intensely with Hardy. He produced a great body of mathematics that today, 60 years later, continues to inspire and challenge number theorists. Some of his formulas even turn out to be important in nuclear physics.

Hardy wrote, "I saw him and talked with him almost every day for several years, and above all I actually collaborated with him. I owe more to him than to anyone else in the world with one exception, and my association with him is the one romantic incident in my life. . . . I liked and admired him . . . a man in whose society one could take pleasure, with whom one could drink tea and discuss politics or mathematics. . . . I can still remember with satisfaction that I could recognize at once what a treasure I had found."[21]

But Hardy also wrote, "Ramanujan was an Indian, and I suppose that it is always a little difficult for an Englishman and an Indian to understand one another properly."

The friendship between the two, while scientifically and intellectually extremely successful and no doubt thrilling to both of them, probably didn't include much real emotional communication. The cold climate, the strange food, the white-faced, distant English people, the wartime conditions, and the separation from his wife and his mother were too much for the young Indian mathematician. He contracted tuberculosis and had to enter a sanatorium. There he must have been even lonelier than before. In January or February of 1918 he threw himself onto the tracks of the London underground in front of an oncoming train. A guard spotted him and brought the train to a screeching stop. Ramanujan was bloodied, and his shins were scarred deeply. He was arrested and hauled to Scotland Yard, and Hardy was called to the scene. He marshaled all his charm and academic

Figure 5-5. Srinivasa Ramanujan, the great Indian number theorist. Courtesy of Archives of the Mathematisches Forschungsinstitut Oberwolfach.

stature and convinced the police that the great Mr. Srinivasa Ramanujan, a Fellow of the Royal Society, simply could not be arrested. "'We in Scotland Yard did not want to spoil his life,' the officer in charge of the case said later."[22]

A few weeks later, Hardy's pretense became true. Back in the sanatorium, Ramanujan received a letter notifying him that he had been elected to the Royal Society. On March 13, 1919, he boarded a ship for home. Back in Madras, he was now an honored celebrity, but his disease continued to worsen, and he died in April 1920 at the age of 34.

It is remarkable how different the tone is in which Hardy writes about Ramanujan. Of Hardy's two great mathematical collaborations, one lasted 35 years, with a fellow Englishman of similar status, education, and culture, and was conducted at arms' length through letters that could be left unanswered even though his collaborator was only a short walk away. The other collaboration, which lasted only a few years, was conducted with a man whose mathematics was intensely interesting to Hardy but much of whose thinking was totally strange to him. That relationship was daily, face to face, with warm personal concern.

Hardy suffered a coronary thrombosis in 1939. In his *Apology*, written a few years later, he wrote:

> I still say to myself when I am depressed, and find myself forced to listen to pompous and tiresome people, "Well, I have done one thing *you* could never have done, and that is to have collaborated with both Littlewood and Ramanujan on something like equal terms." I was at my best at a little past forty, when I was a professor at Oxford. Since then I have suffered from that steady deterioration which is the common fate of elderly men and particularly of elderly mathematicians. A mathematician may still be competent enough at sixty, but it is useless to expect him to have original ideas. It is plain now that my life, for what it is worth, is finished, and that nothing I can do can perceptibly increase or diminish its value.[23]

Snow called Hardy's *Apology* "a passionate lament for creative powers that used to be and that will never come again."[24] Hardy was like "a great athlete, for years in the pride of his youth and skill, so much younger and more joyful than the rest of us, who suddenly has to accept that the gift has gone."[25]

In fact, Hardy, like Ramanujan, attempted to kill himself by swallowing barbiturates, but he took too few. After that, Snow visited him every week. Two or three weeks before his death, the Royal Society informed him that he was to be given their highest honor, the Copley Medal. "He gave his Mephistophelian grin, the first time I had seen it in full splendour in all those months. 'Now I know that I must be pretty near the end. When people hurry up to give you honorific things there is exactly one conclusion to be drawn.'"[26]

Kolmogorov and Aleksandrov

Another remarkable friendship was that between the two famous Russian mathematicians Andrei Nikolaevich Kolmogorov (1903–1987) and Pavel Sergeevich Aleksandrov (1896–1982) Kolmogorov was one of the most original and influential mathematicians of his generation. Aleksandrov was a leading creator of topology and head of the graduate math program at Moscow University during its "golden years" (chapter 6). Shortly before his death, Aleksandrov wrote: "My friendship with Kolmogorov occupies in my life quite an exceptional and unique place: this friendship had lasted for fifty years in 1979, and throughout this half-century it showed no sign of strain and was never accompanied by any quarrel. During this period we had no misunderstanding on questions in any way important to our outlook on life. Even when our views on any subject differed, we treated them with complete understanding and sympathy."[27]

Three years later, in 1986, at the age of 83, Kolmogorov wrote: "Pavel Sergeevich Aleksandrov died six months before my eightieth birthday. . . . [F]or me these 53 years of close and indissoluble friendship were the reason why all my life was on the whole full of happiness, and the basis of that happiness was the unceasing thoughtfulness on the part of Aleksandrov."[28]

The two first met in 1920, but their close friendship began in 1929, at the ages of 26 and 33, when they traveled together for three weeks. In the Caucasus they stayed in an empty monastery on an island in Lake Sevan. Kolmogorov wrote, in his 1986 memoir, "On the island we both set to work. With our manuscripts, typewriter, and folding table we sought out the secluded bays. In the intervals between our studies, we bathed a lot. To study I took refuge in the shade, while Aleksandrov lay for hours in full sunlight wearing only dark glasses and a white panama. He kept his habit of working completely naked under the burning sun well into his old age."[29]

In 1930 and 1931 they traveled in France and Germany. Aleksandrov had been there before, in company with the brilliant young mathematician Pavel Uryson, his close friend and collaborator. Uryson had tragically drowned while swimming off the coast of Brittany. Kolmogorov and Aleksandrov went to Brittany to visit Uryson's grave. "The deserted granite beaches, against which the huge waves thunder, form a complete contrast to the shores of the Mediterranean. Uryson's grave is well tended because it is looked after by Mademoiselle Cornu in whose house Aleksandrov and Uryson were living at the time of his death. Both the gloomy nature of Brittany and the memory of Uryson inclined us to silent walks along the sea shore."[30]

In 1935 Kolmogorov and Aleksandrov bought, from the heirs of the famous actor and director Konstantin Stanislavskii, part of an old manor house in the village of Komarovka. Kolmogorov wrote, "As a rule, of the seven days of the week, four were spent in Komarovka, one of which was devoted entirely to physical recreation—skiing, rowing, long excursions on foot (our long ski tours covered on average about 30 kilometers, rising to 50). On sunny March days we went out on skis wearing nothing but shorts, for as much as four hours at a stretch. . . . Especially did we love swimming in the river just as it began to melt, even when there were still snow drifts on the banks."[31] In Komarovka they were often joined by their students, and they were visited by mathematicians from abroad, including Hadamard and Frechet from Paris, Banach and Kuratowski from Warsaw, and Aleksandrov's collaborator Hopf from Switzerland.

Kolmogorov quotes a letter written on February 20, 1939, from Aleksandrov, at Princeton in the United States, to Kolmogorov in Germany: (They were 43 and 39 years old.) "You have written very little about your sporting activities, but I should like to have a continuous detailed report. . . . Did you swim in the Schwimmhalle? What gymnastics did you do and where? Also you have not written about how you feel. Are you coughing? Are you hoarse? How is your cold? And the main thing, how do you feel in general? It should be a very good idea for you to buy yourself cream as well as milk."[32]

A year after this memoir was published, Kolmogorov was struck by a heavy swinging door and suffered a severe head trauma. As long as he was able, he continued to lecture at the boarding school for talented youngsters that he had founded long before. But the Parkinson's disease, with which he had been afflicted for some time, became much worse. During his last two years he could neither see nor speak. "He died at eighty-four, speechless, blind, and motionless, but surrounded by his students, who for the preceding couple of years had taken turns providing round-the-clock care at his house."[33] Kolmogorov's wife Anna Dmitrievna died only a few months later.

Friends and Colleagues

Many young mathematicians have established friendships and engaged in thoughtful conversations with their professors. Stan Ulam reports:

> Beginning with the third year of studies, most of my mathematical work was really started with conversations with Mazur and Banach. . . . I recall (one such event) with Mazur and Banach at the Scottish Café which lasted 17 hours without interruption except for meals. . . . There would be brief spurts of conversation, a few lines would be written on the table, occasional laughter would come from some of the participants, followed by long periods of silence during which we just drank coffee and stared vacantly at each other. . . . These long sessions in the café were probably unique. Collaboration was on a scale and with an intensity I have never seen surpassed, equaled or approximated anywhere except perhaps in Los Alamos during the war years.[34]

The need to communicate one's insights and discoveries to a friend is present even in those with a less intense emotional bent. The Hungarian mathematician John von Neumann was known above all for his penetrating intellect. Some say that he approached emotional challenges by applying logic to them. But even "Johnny" sought companionship. As a young man, he found it in Eugene Wigner who recalled their intense conversations during their walks: "He loved to talk mathematics—he went on and on and I drank it in."[35]

Von Neumann left Hungary in 1921 for Berlin, Göttingen, and Princeton. His friendship with Wigner continued in Berlin where, being foreigners and not part of the social structure, they became especially close.[36] They reconnected in Princeton and remained friends until von Neumann died of cancer in 1957.

In 1936 Ulam was invited to the Institute for Advanced Study in Princeton by von Neumann. There they became friends. Their conversations were not limited to mathematics, they shared jokes and gossip. The two men had had similar upbringings in wealthy, cultured Jewish homes in Central Europe. Ulam's father was a banker in Lwow, Poland, and von Neumann's was a banker in Budapest, Hungary. During World War II Ulam and von Neumann worked in Los Alamos, and Ulam supported von Neumann's vision of the limitless possibilities of computing. From their free play of ideas came some great advances in applied mathematics: the Monte Carlo method, mathematical experiments on a computer, cellular automata, and simulated growth patterns.

In view of his fantastic mathematical power and his ability to mobilize the diverse conceptual and economic resources that led to the first computer, one might think that von Neumann had enormous self-confidence. But Ulam writes of his friend's self-doubts in his highly informative *Adventures of a Mathematician*. Gian-Carlo Rota also comments: "Like everyone who works with abstractions, von Neumann needed constant reassurance against deep-seated and recurring self-doubts."[37]

Another Hungarian mathematician, Paul Erdős, is legendary for the number and importance of his collaborations. In the 1920s he led a group of young men and women, the self-named "Anonymous Group," who met weekly to explore questions in discrete mathematics. (We write about this group in chapter 6.) Long after the Anonymous Group broke up, Erdős continued to find collaborators everywhere. He published all together about

Figure 5-6. Paul Erdős (right), Hungarian mathematician, with one of his friends, Aryeh Dvoretsky. Courtesy of Archives of the Mathematisches Forschungsinstitut Oberwolfach.

1500 papers, second only to the immortal Leonhard Euler. It's claimed that once, when Erdős had to take a long train ride, he wrote a joint paper with the train conductor!

There is a number called the "Erdős number," and every mathematician has one. If you collaborated directly with Erdős, your Erdős number is 1; if you never collaborated directly with him but did collaborate with one of his collaborators, your Erdős number is 2. And so on. If you're a well-known mathematician active during Erdős' career, your Erdős number is almost surely less than 10. And if you don't have a finite Erdős number, your Erdős number is infinity. Erdős himself, of course, had Erdős number zero.

Erdős' collaborator often had to fill in details and write up results for publication. Usually Erdős had created the problem. If you were lucky enough to be one of his many mathematical friends, from time to time he would show up at your home, announcing, "My brain is open!" For the next few days, or maybe a week or two, you or your spouse had the privilege of providing him with food, lodging, chauffeuring and long-distance telephone service. Then he would move on. If he stayed longer than you could tolerate, it was all right to ask him to leave.

Once at Stanford University he lived with his friend Gabor Szegő and gave no sign of imminent departure. One night Szegő's wife met their friend Andras Vázsonyi at a party and said, "Erdős dropped in three weeks ago and he is still staying with us. I am at the end of my wits." Vázsonyi told her, "No problem. Tell him to get out." "I couldn't do that," she said. "We love him and could not insult him." "Do what I said," Vázsonyi insisted. "He will not be insulted at all." An hour later Erdős came up to Vázsonyi and asked for a ride to a hotel. "What happened?" Vázsonyi asked innocently. "Oh, Mrs. Szegő asked me to move out because I stayed long enough," he said, totally undisturbed."[38]

Erdős had a famous collaboration with the Polish probabilist Mark Kac. Erdős happened to be in the audience at the Institute for Advanced Study in Princeton when Kac gave a talk revealing the amazing fact that the number of prime factors of a random integer is distributed like a bell curve. Unfortunately, Kac was still lacking one difficult estimate to clinch the proof. At that time Erdős knew nothing about probability, so he had been dozing, but he woke up when Kac said "prime divisor." Before the talk was over, he showed Kac the missing proof.

Something similar happened with the Norwegian number theorist Atle Selberg, but with a less happy ending. Selberg had a conversation with Erdős at a moment when he was in hot pursuit of the long missing "elementary" proof of the prime number theorem (the one that says how primes are distributed asymptotically and logarithmically). Erdős quickly supplied a proof to fill in a difficult gap that Selberg was confronting. Unfortunately, word quickly got around, and in a few days Selberg was told the exciting news that "Erdős and some Scandinavian mathematician" had found the missing elementary proof. Selberg was so offended that a permanent breach was created between them.

This was all the more painful to Erdős since he never begrudged sharing ideas or credit.

Erdős is often written about as an odd eccentric. But he was more an object of love than of laughter. It is true that he had no permanent address or job; he carried all his needs in a half-full suitcase. For much of his life, his mother was a constant companion. When he left Hungary for Britain, in his early twenties, it seemed that he had never before had to butter his own bread. But for those who knew him, in his long and extremely productive life, his outstanding characteristic was his kindness and selflessness. He naively and innocently expected others to do much for him; but he was always ready to share whatever he had with anyone who needed or could use his help. When he won a prize, he immediately gave the money away to other mathematicians who needed it more than he.

Gödel and Einstein

Another famous friendship, and one of the most unusual, was between the mathematical logician Kurt Gödel, then in his thirties, and Albert Einstein, the world-renowned physicist, in his seventies. Gödel is famous for his "incompleteness theorem": Any formal language and system of formal axioms, which are strong enough to generate the natural number system, necessarily will also include an "undecidable" sentence or formula, one that the axioms can neither prove nor disprove.

Gödel and Einstein were often seen walking together in Princeton, deep in German conversation. Although they were very different, they "appreciated each other enormously."[39] Einstein was friendly, frequently joyous, while Gödel was fearful, hard to relate to, and had experienced several episodes of depression. In spite of these differences, the two men wrote enthusiastically about the value of their connection. Gödel wrote to his mother that "there was simply nobody else in the world with whom to

Figure 5-7. Albert Einstein and Kurt Gödel, friends at Princeton. Courtesy of The Shelby White and Leon Levy Archives Center, Institute for Advanced Study, Princeton, NJ, USA. Gift of Dorothy Morgenstern Thomas.

talk, at least not in the way I could talk to Einstein."[40] Many wondered about this unusual relationship, but to the novelist Rebecca Goldstein it is understandable: "Strange as it might seem for men as celebrated for their contributions, they were intellectual exiles. . . . I believe they were fellow exiles in the deepest sense in which it is possible for a thinker to be an exile."[41] Einstein was self-exiled by his persistent seeking for a unified field theory for many years, when quantum theory was the center of interest for theoretical physics, and Gödel was self-isolated from virtually all the rest of the world.

Einstein died first. Gödel's only remaining regular connection was with his wife Adele. When Adele became ill and they had to be separated, he thought any food not prepared by Adele might be meant to poison him. The fatal consequence was described in the last chapter.

Mathematical Marriages

We will mention two examples of mutual support and love among mathematical married couples. Our first example is the remarkable mathematical partnership of Grace Emily Chisholm (1868–1944) and William Henry Young (1863–1944). Grace and Will left behind a mass of documents and letters that Ivor Grattan-Guinness used to write "A Mathematical Union."[42] His article, in which he quotes many of the Youngs' letters, is the source of this account.

William Young won a mathematics scholarship to Cambridge at age 16. After graduation he spent his spare time and money on sports, particularly rowing. In 1888 he was appointed lecturer in mathematics at the women's college, Girton. While working there he managed to save 6000 pounds for a future of leisurely travel. "But these plans suffered an abrupt halt, for early in 1896 he fell in love with Grace Emily Chisholm."[43]

Grace had passed the Cambridge senior examination in 1885 at the age of 17. Her family encouraged her to devote herself to charitable duties, such as visiting the poorest parts of London where the police dared to venture only in pairs. Nevertheless she studied mathematics and won a prestigious scholarship at Girton College, Cambridge. But she was soon disenchanted. She wrote that at Cambridge it was believed that "mathematics had reached the acme of perfection, with nothing left to do but fill in the details." Of Professor Arthur Cayley, the famous reigning Cambridge mathematician, she wrote, "Cayley sat, like a figure of Buddha on its pedestal, dead weight on the mathematical school of Cambridge." Cayley's lectures were "a flow of words. . . . Polyhedra with vertices constantly springing from triangular faces, like crystals growing in a solution, trees with branches forking in all directions succeeded one another without intermission, twining this way and that round the professorial head, or emerging from under his flapping sleeves as he stood with his back to the listeners chalking and talking at the same time at the blackboard."[44]

Figure 5-8. Grace Chisholm Young. Source: *More Mathematical People: Contemporary Conversations.* Eds. Donald J. Albers, Gerald L. Alexanderson, and Constance Reid. Figure 5-9. William Young, British analyst, husband of Grace.

Grace heard about Mr. Young. As a tutor, he had a reputation for cramming his students through examinations, and making his girl students cry. So she decided definitely to go not to him but to Mr. Berry at King's College. But in her third year, Mr. Berry took a term off and Grace was assigned William Young as her temporary tutor.

She first saw him while sitting on a window seat in a friend's room. "This dreaded Mr. Young, this great mathematical gun, was a boy, hardly older than herself. The face, shaded by a simple sailor hat with his college colours, was a fine one, with a marble look, due to a pure line of feature and the clear delicacy of skin. . . . The fixed look in the eyes and the slight speaking movement of the mouth made evident that he was working out some mathematical problem as he passed and disappeared under the archway."[45]

With coaching from Will, Grace graduated with honors from Cambridge in 1892, and then went to Göttingen for graduate study. She wrote home about her professor Felix Klein who was sensitive to the presence of a female student: ". . . Instead of beginning with his usual 'Gentlemen!' [he] began 'Listeners!' ('*Meine Zuhörer*') with a quaint smile; he forgot once or twice and dropped into 'Gentlemen' again, but afterwards he corrected himself with another smile. He has the frankest, pleasantest smile, and his whole face lifts up with it."[46]

In 1895 Grace obtained her Ph.D. by applying Klein's group theory to spherical trigonometry. This was a tremendous triumph, for it was the first doctorate awarded to a woman in Germany in any subject whatever. Sonya Kovalevskaya's doctorate had been given unofficially, as she had not taken courses at Göttingen nor taken a *viva voce* exam.

Early in February Will proposed to Grace. "When Grace told him that she could neither marry him nor anyone else, he did not hear a word. . . . She soon fell deeply in love with him. She never dared disillusion him about their betrothal."[47] But now Will began to think of research for the first time. He believed he would not be able to match the abilities of his talented wife, but at least he might be able to make some contribution of his own.

A year after the wedding, their first child, Frank, arrived, and they decided to move to Göttingen. Grace wrote, "At Cambridge the pursuit of pure learning was impossible. . . . Everything pointed to examination, everything was judged by examination standard. There was no interchange of ideas, there was no encouragement, there was no generosity."[48] Her husband, like most of the other fellows, worked hard to support his family by taking work at Girton and Newnham and by presiding at local examinations. Such work allowed him to provide comforts and luxury for his wife, which he considered important. But Chisolm rejected such a view and saw it as undermining the values of an older English society.

At Göttingen, under Klein's encouragement, Will wrote his first original paper, on problems in geometry. Research was now fully joined for both of them—for him, in his 35th year.

After a few years, money needs forced Will to go back to Peterhouse at Cambridge while Grace and their child stayed in Göttingen. This would be their pattern of life for many years, Will traveling to and from the family home, Grace bringing up the family and following her other interests, and both of them working intensely on mathematics.

In 1900 Klein suggested that they read two articles by Arthur Schoenflies on Georg Cantor's revolutionary theory of infinite sets. This was excellent advice. Set theory, and its application to mathematical analysis, was the field where Grace and Will would work for the next 25 years.

> And as their work developed, an extraordinary reversal took place. . . . Will had a profound and original mathematical mind, and, in the field in which he concentrated, one of the finest minds in the world. . . . Grace became his secretary and assistant, perfectly capable of making original contributions of her own but basically needed to see that the flood of ideas that was poured out to her could actually be refined into rigorous theorems and results . . . the success of Will's late start was due to the support he received from his talented wife. During their twenty-five years of mathematical research, they published between them three books and about 250 papers. When they were apart, their letters constantly discussed mathematical questions: when they were together, their conversation was so dominated."[49]

In a letter Will justified his frequent publication as sole author of their joint work. "Our papers ought to be published under our joint names, but if this were done neither of us get the benefit of it. Mine the laurels now and the knowledge. Yours the knowledge only. Everything under my name now, and later when the loaves and fishes are no more procurable in that way, everything or much under your name. . . . But we must flood the societies with papers. They need not all of them be up to the continental standard, but they must show knowledge which others have not got and they must be numerous."[50]

By 1904 Will had independently constructed his own theory of integration equivalent to Lebesgue's, which had been published earlier. The Lebesgue integral is one of the keystones of functional analysis. Young's approach was significant at the time and was preferred by some later writers. Just before Easter 1907, Will was elected a Fellow of the Royal Society.

In February 1900 their second child, Rosalind Cecilia Hildegard, was born. In December 1901 the third arrived, Janet Dorothea Ernestine, and in September 1903, the fourth, Helen Marian Kinnear. In July 1904 Grace gave birth to Laurence, who would grow up to be a mathematician and a long-time professor at the University of Wisconsin. Continuing the family tradition, Laurence's daughter Sylvia Wiegand became an algebraist at the University of Nebraska and was president of the Association for Women in Mathematics.

The Young's last child, Patrick Chisholm, was born in March 1908. That year they moved from Göttingen to Switzerland. In 1908 Will published nearly 20 papers; and in 1910, 22. But his applications for professorships were rejected by Liverpool in 1909, by Durham and Cambridge in 1910, and by Edinburgh and King's College London in 1912.

In his joint biography Grattan-Guinness writes, "It seems unbelievable that a man who was producing so much profound research could not obtain a post in his own land . . . his career had been unforgivably unconventional: no regular appointments after the days of Cambridge coaching, living abroad during middle life, a sudden burst of original work after years of silence. Will had stepped out of line, and he was not allowed to step back into it again. . . . But Will was not the easiest of people to live with or to know. His letters show his impatience and over-sensitivity, and his desire to impose his view upon others."[51]

Finally, in August 1913, Will obtained the Hardinge Professorship of Mathematics at the University of Calcutta, at 1000 pounds per year plus expenses for a few months' attendance during the year. But Will wrote to Grace, "The sun is a fierce enemy. . . . What with malaria, smallpox, typhoid and sun stroke one ought to be very heavily bribed if one is to come at all."[52]

Now Grace started to write papers under her own name again. She wrote a series of papers on the foundations of the differential calculus, starting auspiciously with a paper in 1914 in the Swedish journal *Acta Mathematica* and reaching a peak with a long essay that won the Gamble Prize at Cambridge in 1915.

"When Will was at home he completely monopolized Grace's life and duties. He could not help himself and realized that one of the advantages of his travels was that it would give Grace a period of quiet and undisturbed work."[53]

In 1914 the Great War came. Their oldest son Frankie interrupted his engineering course at Lausanne to volunteer for the Royal Flying Corps. On February 14, 1917, at 5000 feet Frank's plane was attacked by nine German fighters: he was shot through the head and died.[54]

Grace later wrote, "It was Sunday morning, February 18th, when the doorbell rang. Ah! That bell at an unexpected time, there was one thing it might always mean. Too true, there stood the postman with a telegram. I tore it open—War Office. . . . You hundreds of thousands who have gone through what we went through, you will have a vision of those awful hours of agony!"[55] Grace and Will heard from Émile Picard and Jacques Hadamard, who had each lost two sons; from Emile Borel, who had lost his adopted son; from E.W. Hobson, whose youngest son had had a mental breakdown; and from George Polya, who had lost his brother.

Will continued to work and during 1916 and 1917 published over 20 papers. During the summer he received the de Morgan Medal from the London Mathematical Society, an award made every 3 years for distinguished contributions to mathematics. In 1919 he became professor at the University of Wales at Aberystwyth. He was in his middle fifties. He was later described by Sir Graham Sutton, who came up to Aberystwyth in the autumn of 1920: "He was a tall, vigorous man with the most immense beard I have ever seen. One's immediate impression was he was bursting with energy. He did all things quickly, was given to forthright speech, but when he wished he could be the most persuasive and charming of men . . . he made mathematics exciting. It was one of the more memorable experiences of my life to sit at the feet of one who knew so much mathematics."[56]

Will served a term as president of the London Mathematical Society. In his retirement speech in November 1922, he also announced his retirement from active mathematical research. Not only were his own powers declining, but Grace could no longer keep up her role in their partnership. He ended his address with a quote from Prospero's farewell speech in *The Tempest* "This rough magic I here abjure . . . I'll break my staff, bury it certain fathoms in the earth, and deeper than did ever plummet sound, I'll drown my book."[57]

For Grace his retirement meant freedom to achieve something substantial of her own. She was troubled by gallstones and much taken up with housekeeping—running the gardens and the vineyards, making jam and wine—yet in 1929 she began *The Crown of England*, a historical novel in the style of Sir Walter Scott. It required an enormous amount of research from original sources, took 5 years to complete, and was nearly 400 pages long, together with a number of drawings which she prepared herself. Will took the book to several publishers during his visits to London, but none of them accepted it.

In 1929 Will became president of the International Mathematical Union. He had ambitious plans for reforming international mathematical organization, but despite much effort nothing came of it. "Disappointment and disillusion loomed large in his life from this time onwards."[58]

When World War II broke out, Grace was in England. To return to Will in Switzerland would have required a dangerous sea voyage to Spain, and after that a hazardous crossing through Spain and the unoccupied parts of France. Her health could not take the strain. Will became senile, was moved into a nursing home, and died on July 7, 1942, a few months before his 79th birthday. For Grace, his death was a relief. "That is the solution."[59] In March 1944 she was proposed for the rare award of Honorary Fellow of Girton College; but before it could be approved she suffered a heart attack and died on the evening of the 19th, a fortnight after her 76th birthday.

Julia and Raphael Robinson

The marriage between Grace Chisholm and Will Young was controlled and shaped by the relation between the sexes as it was in the early 20th century, before the Great War, as they then called it, and between World Wars. The changes in the relations between men and women in recent decades are illuminated by the story of the marriage of Julia Bowman and Raphael Robinson, as told by Julia's sister, the author Constance Reid, in a short book, *Julia*. Constance's biography of her sister is written in the first person. As Constance was writing it in 1985, Julia was dying of leukemia, at the age of 65. But Julia did hear and approve of everything that Constance wrote.

Julia Robinson became famous for her key role in solving the 10th in the list of 23 problems that David Hilbert proposed to the 1900 International Congress. The 10th problem is: to give a method or algorithm, to determine whether an arbitrary polynomial equation with integer coefficients has an integer solution. (This is called a "Diophantine equation".) Together with Martin Davis and Hilary Putnam of the United States, Julia Robinson achieved partial results which, when completed by the young Russian Yuri Matyasevich, proved that no such algorithm exists.

Julia's life as a mathematician was closely bound up with her husband Raphael, a professor of mathematics at Berkeley. As a child, Julia contracted scarlet fever, then rheumatic fever, and was bedridden for nearly two years. Unknown to her at the time, she suffered heart damage that would handicap her for life. These experiences taught her patience (an important trait for mathematicians). In high school she took mathematics classes, where frequently she was the only girl. Then at San Diego State College she wanted to become a mathematician, but the course offerings were limited. (This was before that college became a University.) Not until she became a graduate student at Berkeley did Julia Bowman find the challenge and stimulation she needed.

We quote from Reid's book as it is written, in the first person of Julia Robinson:

> I was very happy, really blissfully happy, at Berkeley. Mine is the story of the ugly duckling. In San Diego there had been no one at all like me. . . . Suddenly at Berkeley, I found that I was really a swan. . . . Without question what had the greatest mathematical impact on me at Berkeley was the one-to-one teaching that I received from Raphael.[60] During our increasingly frequent walks, he told me about various interesting things in mathematics. He is, in my opinion, a very good teacher. I doubt that I would have become a mathematician if it hadn't been for Raphael. He has continued to teach me, has encouraged me, and has supported me, in many ways, including financially.[61] . . . I was offered a job as a night clerk in Washington, D C at $1200 a year. My mother thought that I should accept it, but Raphael had other ideas. . . a few weeks after the Japanese attacked Pearl Harbor, Raphael and I were married."[62]

At mid-century, nepotism rules were still in effect. A husband and wife couldn't both be members of the Berkeley mathematics faculty. "Because of the nepotism rule I could not teach in the mathematics department, but this fact did not particularly concern me. Now that I was married, I expected and very much wanted to have a family."[63] Julia did become pregnant, but had a miscarriage. Then she contracted viral pneumonia. Her doctor discovered that she had a serious heart ailment and advised her under no circumstances to become pregnant again. He told her mother that she would probably be dead by forty, since by that time her heart would have broken down completely.

Figure 5-10. Raphael Robinson and Julia Robinson. Courtesy of Dolph Briscoe Center for American History, The University of Texas at Austin. Identifier: di_05556. Title: Rapfael and Julia Robinson. Date: 1978/04. Source: Halmos (Paul) Photograph Collection

"For a long time I was deeply depressed by the fact that we could not have children. Finally Raphael reminded me that there was still mathematics."[64] She began to work toward a Ph. D. with Alfred Tarski., the great logician who led the logic group at Berkeley. She did not receive a regular appointment at Berkeley until the announcement of her election to the National Academy of Sciences in 1976. When the university press office received the news, someone there called the mathematics department to find out who Julia Robinson was. "Why, that's Professor Robinson's wife." "Well," replied the caller, "Professor Robinson's wife has just been elected to the National Academy of Sciences."[65]

"The University offered me a full professorship with the duty of teaching one-fourth time—which I accepted."[66] In 1982 Julia was nominated for the presidency of the American Mathematical Society. "Raphael thought I should decline and save my energy for mathematics, but I decided that as a woman and a mathematician I had no alterative but to accept. I am planning to take his work as the subject of my Presidential Address at the AMS Meting in New Orleans this winter."[67]

Friendships Among Women Mathematicians

There are few records available about women and their connections to other women in mathematics. Until the impact of the women's movement, women mathematicians were usually surrounded by men in their classes and at their jobs. A high percentage of these women were (and are) married to mathematicians, whom they frequently rely on for discussion of their work and for support.

As the participation of women increased in the mathematical profession, more friendships between them were established. In an autobiographical essay written late in her life, Olga Taussky-Todd recalls her childhood in Olmutz, Austria (now in the Czech Republic). As a child, she had strong relationships with young women with intellectual interests, including her sisters. But once she entered the University of Vienna, all her contacts were with male mathematicians. Her account of those years describes her teachers, her research, and her presentations, one of which led to a temporary appointment in Göttingen. Her task there was to edit Hilbert's work on number theory. It was then that she met Emmy Noether, whom she admired greatly. Noether was one of the most important mathematicians of the 20th century, the principal creator of modern abstract algebra. As a woman, a pacifist, and a Jew, she was not considered fit for a professorship in Germany and remained an unpaid adjunct to the Göttingen faculty through her most important creative years.

Taussky-Todd writes, "I had the good fortune to gain her confidence through an act of concern for her that had seemed very natural to me, and we became good friends. I had been present when one of the top people of the department spoke rather harshly to her. I really did not like this. The next day I told him that this had upset me. . . . He went to apologize to Miss Noether."[68]

After both of them left Göttingen, they met at Bryn Mawr College and began spending more time together. The appeal of Bryn Mawr was due to the reputation of Anna Pell Wheeler, chair of the mathematics department and a strong advocate for women in the field. In inviting Emmy Noether to her college, Wheeler gave the program a visibility which made this department renowned for decades. By the time Olga Taussky came to the United States in 1934, Noether was ill. But she tried to hide this fact from her colleagues and students. She continued to teach and to travel to Princeton, and Taussky frequently went with her. These trips to Princeton were the highlights of her year at Bryn Mawr.[69] Her characterization of Noether evokes a complex woman: generous, brilliant, devoted to her friends and students, but also a person with limited interpersonal skills, who needed the kind of support Taussky provided.

Subsequently, Taussky returned to England where she obtained a fellowship at Cambridge University. There she met her future husband, Jack Todd, also a mathematician, with whom she had a deep emotional and intellectual bond. They worked together in applied areas during World War II. In her autobiographical essay, Taussky-Todd recounts an extraordinary number of stimulating and productive interactions with mathematicians throughout her career; but none had the same impact upon her as her friendship with Emmy Noether.

Their relationship was quite rare as illustrated by the findings of the psychologist Ravenna Helson who studied women mathematicians in the 1950s. She interviewed women who were identified by their peers as creative and found that they did not have as many friends and collaborators as their male counterparts. From this finding (which was surprising as women generally report more interpersonal ties than men do), two issues arose. One of these was that women in the 1950s had very few

Figure 5-11. Olga Taussky-Todd, Austrian-American algebraist and number theorist. Courtesy of Archives of the Mathematisches Forschungsinstitut Oberwolfach

female colleagues. For instance, Vera Pless, a mathematician at the University of Illinois, wrote that she never saw a female peer the entire time she was a student.[70] While she did know about the work of Emmy Noether (which may have influenced her choice of algebra as her area of specialization), her actual contacts were with men. The second issue is that of time. In Helson's interviews, "several mentioned being overworked and underpaid. One participant talked about the under-stimulation and lack of time to herself that she experienced at a women's college; so she moved to a university where she loved the library and the intellectual stimulation but she was not promoted at the same rate as her male colleagues who had published less."[71]

The experience of African-American women mathematicians reflects the complexities of discrimination and mutual support. The theme of minority mathematicians and women runs through a number of chapters of this book. We first mentioned Vivienne Malone-Mayes in chapter 1, and we devote chapter 8 to issues of race.

Malone-Mayes was a graduate student at the University of Texas in the early 1960s. She recalled her experiences in that institution in a panel presentation: "My personal isolation at the University of Texas in Austin was absolute and complete, especially during the summer of 1961. At times I felt that I might as well have been taking a correspondence course. For those who completed degree programs, and for many who quit along the way, the lack of interchange with fellow students was a profound hindrance to academic achievement."[72]

Various universities treated women differently. As mentioned in chapter 1, the Courant Institute at New York University had a number of faculty members at midcentury who encouraged and supported women mathematicians.

Joan S. Birman, currently a professor at Columbia University, started graduate studies later in life. While at the Courant Institute, she found that her male and female fellow students were open to interacting with her, and that they were helpful to each other while working on their dissertations. This early experience encouraged her to continue to collaborate actively with colleagues throughout her career. She writes, "I've often wondered whether, if the mathematical community welcomed older women as graduate students in a serious and non-patronizing way, and if women rejected the myth that mathematics was a young man's game, we might see real changes in those discouragingly low numbers."[73]

At the University of Michigan in the early 1970s, African-American students were admitted in greater numbers than elsewhere. Janice Brown Walker's description of her graduate experience was positive. She wrote, "[In fall 1971] I was relieved and excited to see more than six other African-American graduate students there . . . [who] formed a closely knit group that still exists. We were a family. We celebrated successes and shared failures . . . [this group] also formed the core of a mathematical society that was organized as a forum for providing support and information to each other, presenting mathematical talks to each other, and interacting socially. . . . The sheer number of us attending made it easier to develop a group sense of power, courage, and self-esteem. Also, we were warmly accepted and supported by a number of [graduate students and] faculty members.[74]

The power of friendships is in the creation of an environment where new ideas can be explored before they are open to the more rigorous, critical scrutiny of the larger, professional community. The psychologist Howard Gardner writes that creative breakthroughs are sustained by friends, peers, and partners in two ways: emotionally, "in which the creator is buoyed with unconditional support," and cognitively, "where the supporter seeks to understand, and to provide useful feedback on, the nature of the breakthrough."[75]

Bibliography

Abelson, P. (1965). Relation of group activity to creativity in science. *Daedalus* (summer), p. 607.

Albers, D. J., & Alexanderson, G. L. (Eds.) (1985). *Mathematical people: Profiles and Interviews*. Boston: Birkhäuser.

Aleksandrov, P. S. (2000). "A few words on A. N. Kolmogorov," in *Kolmogorov in Perspective*. Providence, R.I.: American Mathematical Society, pp. 141–144.

Beaulieu, L. (1993). A Parisian café and ten proto-Bourbaki meetings (1934–1935). *Mathematical Intelligencer* 15(1) 27–35.

Birman, J., Haimo, D. T., Landau, S., Srinivasan, B., Pless, V., & Taylor, J. E. (1991). In her own words. *Notices of the American Mathematical Society*, 38(7), 702–706.

Bollobás, B. (Ed.). (1986). *Littlewood's miscellany*. Cambridge: Cambridge University Press, p. 8.

Brown-Walker, J. (2005). A double dose of discrimination. In B. A. Case & A. M. Leggett (Eds.). *Complexities: Women in mathematics*. Princeton, N.J.: Princeton University Press, pp.189–190.

Cartan, H. M. (1980). Nicolas Bourbaki and contemporary mathematics. *Mathematical Intelligencer* 2(4), 175–187.

Case, B. A., & Leggett, A. M. (Eds.). (2005). *Complexities: Women in mathematics.* Princeton, N.J.: Princeton University Press.

Fleck, L. (1979). *Genesis and development of a scientific fact.* Chicago: University of Chicago Press.

Gardner, H. (1993). *Creating minds.* New York: Basic Books.

Gessen, M. (2009). *Perfect rigor.* Boston: Houghton Mifflin Harcourt.

Goldstein, R. (2006). *Incompleteness: The proof and paradox of Kurt Gödel (Great Discoveries).* New York: W. W. Norton.

Grattan-Guinness, I. (1972). A mathematical union, *Annals of Science 29*(2), 105–186.

Hardy, G. H. (1967). *A mathematician's apology.* New York: Cambridge University Press.

Hardy, G. H. (1978). *Ramanujan.* New York: Cambridge University Press.

Heims, S. J. (1980). *John von Neumann and Norbert Wiener: From mathematics to the technologies of life and death.* Cambridge, Mass.: MIT Press.

Helson, R. (2005). Personal communication.

James, I. (2002). *Notable mathematicians.* New York: Cambridge University Press.

Kanigel, R. (1991). *The man who knew infinity.* New York: Simon and Schuster.

Koblitz, A. (1983). *A convergence of lives.* Boston: Birkhäuser.

Kolmogorov, A. N. (2000). "Memories of P. S. Aleksandrov," in *Kolmogorov in perspective.* Providence, R.I.: American Mathematical Society.

McMurran, S. L., & Tattersall, J. J. (1996). The Mathematical collaboration of M. L. Cartwright and J. E. Littlewood, *American Mathematical Monthly 103*(10), 833–845.

Reid, C. (1996). *Julia: A life in mathematics.* Washington, D.C.: MAA Spectrum.

Reid, C. (2004). *Hilbert.* New York: Springer.

Rota, G. C. (1987). The lost café. *Los Alamos Science*, Special Issue, p. 26.

Schechter, B. (1998). *My brain is open: The mathematical journeys of Paul Erdős*. New York: Simon and Schuster.

Snow, C. P. (1993). *The two cultures*. Cambridge: Cambridge University Press.

Szanton, A. (1992). *The recollections of Eugene P. Wigner as told to Andrew Szanton*. New York: Plenum Press.

Tattersall, J., & S. McMurran (2001). An Interview with Dame Mary L. Cartwright, D.B.E., F.R.S. *College Mathematics Journal* 32(4) 242–254.

Tausky-Todd, O. (1985). Autobiographical Essay. In D. J. Albers and G. L. Alexanderson (Eds.). *Mathematical people: Profiles and interviews*. Boston: Birkhäuser.

Ulam, S. (1976). *Adventures of a mathematician*. New York: Scribners.

www.maths.abdn.ac.uk/courses/mx4531/chap_gamma/pdf/ram .pdf

+6+

Mathematical Communities

What sorts of communities do mathematicians form? How do their communities shape their lives? We will describe some informal groups, which were formed around specific needs of their participants. Whether inside or outside universities, communities that were fueled by a shared vision have brought about significant change in mathematics.

Insiders and Outsiders

Stan Ulam has written:

> Much of the historical development of mathematics has taken place in specific centers. These centers, large or small, have formed around a single person or a few individuals, and sometimes as a result of the work of a number of people—a group in which mathematical activity flourished. Such a group possesses more than just a community of interests; it has a definite mood and character in both the choice of interests and the method of thought. . . . The great nineteenth-century centers such as Göttingen, Paris and Cambridge (England) all exercised their own peculiar influence on the development of mathematics.[1]

In this chapter we will look at three 20th century communities based in university mathematics departments: Göttingen in Germany, from the 1890s to the 1930s; its New York offspring, the Courant Institute, starting in the 1930s; and the Department of Mechanics and Mathematics (Mekh-Mat) at Moscow State University in its "golden age" of the 1960s. Göttingen and Moscow are tragic examples, showing the fragility of intellectual communities under hostile regimes.

We also look at four communities that are intentional rather than organizational, established by people with a strong common goal or interest who did not find any existing organization to serve their purpose. One example was the Jewish People's University in Moscow from 1978 to 1983. Another was the Anonymous Group in Budapest, led by Paul Erdős. A more famous example was the French group Bourbaki. Originally inspired by a desire to renew and modernize French mathematics, it later focused on producing a series of texts, held together by a common vision of what mathematics could or should be (formal, axiomatic, abstract) and a rebellious or combative attitude toward the classical French mathematical program of analytic function theory.

Our last example is a contemporary group, the Association for Women in Mathematics (AWM). It provides a meeting place for female mathematicians and students, whose members work and fight to improve their status and recognition. (Membership is not restricted to women and there are male members as well.) It is a community of practice and belief. It has a lot in common with the National Association of Mathematicians (NAM), which addresses the needs and concerns of African-American mathematicians and mathematics students. There is also a group of Chicano and Native American mathematicians—The Society of Chicanos and Native Americans in Science (SACNAS)—but it is not devoted solely to mathematics. These contemporary communities are greatly assisted by electronic communication, as well as face-to-face meetings, and are comparable to the Anonymous Group and Bourbaki in being autonomous, self-created communities based on a common interest.

Clearly there are many overlapping interacting mathematical communities of various sizes, kinds, lifetimes. There are research communities, there are publishing communities, there are teaching communities, and there are even bureaucratic communities (for example, the groups of mathematicians associated with the National Science Foundation). One can think of the whole mathematical community as the union of all these smaller subcommunities.

Active researchers are always to some degree members of communities in certain subject areas. Some researchers, like Kurt Gödel and Andrew Wiles, have been private, even secretive for a while; others, like Bill Thurston and Paul Erdős, have been outgoing and communicative; but in either case, the research is motivated and ultimately evaluated by a community of some kind, whether face to face or electronic. The membership list in such a community is less definite, membership is a matter of degree; it may be variable or even controversial.

The historian David Rowe has written:

> A fundamentally new type of mathematical community has now rendered the traditional nineteenth-century modes of communication and invention largely obsolete. Mathematics today is essentially an oral culture; to keep abreast of it one must attend conferences and workshops, or, better yet, be associated with a leading research center where the latest developments from near and afar are constantly being discussed. By the time an important result actually appears in print today, it is probably no longer new; in any case, it will probably be impossible to understand the work without the aid of an "intervener" who already knows the thrust of the argument through an oral source.[2]

In the 20 years since Rowe wrote this, e-mail networks have again transformed mathematical communication and made it much more rapid. His point about the need to be "plugged into the loop" in order to keep up with current research is much more valid today.

At the American Institute of Mathematics in Palo Alto, California, focused workshops are held throughout the year. Several mathematicians interested in a common problem are invited to meet each other there and spend a week in concentrated joint effort.

A totally new form of mathematical collaboration was started in February 2009 by Timothy Gowers of Cambridge University in England, whose work in combinatorics has won him a Fields Medal. He called it the Polymath Project. A carefully chosen problem was proposed for public access on the Web: "to find an elementary proof of a special case of the density Hales-Jewett theorem (DHJ), which is a central result of combinatorics." Anyone who wanted to join in could post suggestions or calculations toward solving the problem. The results were indeed impressive. A major open problem was solved in a few weeks by the shared effort of over two dozen contributors from several countries. Gowers wrote:

> When the collaborative discussion kicked off on February 1, it started slowly; more than seven hours passed before Jozsef Solymosi, a mathematician at the University of British Columbia in Vancouver, made the first comment. Fifteen minutes later a comment came in from Arizona-based high-school teacher Jason Dyer. Three minutes after that Terence Tao (winner of a Fields Medal, the highest honour in mathematics) at the University of California, Los Angeles, made a comment. Over the next 37 days, 27 people contributed approximately 800 substantive comments, containing 170,000 words. . . . Progress came far faster than anyone expected. On March 10, Gowers announced that he was confident that the Polymath participants had found an elementary proof of the special case of DHJ, but also that, very surprisingly, the argument could be straightforwardly generalized to prove the full theorem.[3]

Where is the emotional side of this? Any community has members and nonmembers. To be a community means to include and to exclude. Membership provides rights and privileges. Exclusion withholds some or all of those rights and privileges. There might be some persons who would like to be included but who are excluded. So, of course, inclusion or exclusion from a community has an emotional side. Inclusion brings security and solidarity. Exclusion, for good reason or bad, may provoke resentment and hostility.

Ideally, access to the mathematics community would be based solely on mathematical merit. If you are a creative mathematician, if you solve hard problems or invent interesting concepts, you should be welcome. And by and large, that is pretty much true.

On the other hand, things aren't always that simple. The mathematical community never did exist in a vacuum. Somebody has to pay the bills. The money has to come from somewhere. As we will see in this chapter, attached to the money may be values and prejudices—political prejudices, nationalist prejudices, religious prejudices, race prejudices, gender prejudices, and age prejudices. The price paid for being female by Sophie Germain, Sonia Kovalevskaya, and Emmy Noether is discussed in the next chapter. Issues of race, language, nationality, and ideology are also relevant, as during the Cold War. Age can be a problem too, if one is too young or too old (see chapter 7).

Some people have found entry into the mathematical community a challenging and at times discouraging process. The famous mathematical statistician Herbert Robbins was asked, "Were there any mathematicians who gave you guidance and encouragement during critical periods of your professional development?" He answered, "No. What they gave me was something perhaps more important. The leading mathematicians I encountered made me want to tell them, 'You son-of-a-bitch, you think that you're smart and I'm dumb. I'll show you that I can do it too!' It was like being the new kid in the neighborhood. You go out into the street and the first guy you meet walks up to you and knocks you down. Well, that's not exactly guidance or encouragement. But it has an effect." But in the same interview Robbins also said, "Marston Morse impressed me deeply. I could see that he was on fire with creation. . . . He was, in a way, the type of person I would like to have been."[4]

Some stories of pathos, incomplete promise, and exclusion are famous among mathematicians and have been told in Eric Temple Bell's much-read *Men of Mathematics*. Evariste Galois (1811–1832) died before he was 21 in a stupid, meaningless duel. He had already been noticed and marked as a brilliant young mathematician. On his own, he had penetrated to the heart of the theory of polynomials—the algebra of permutations of the roots, which we now call Galois theory. But this was at the time of the restoration of the Bourbon monarchy in France, which followed the French Revolution in the late 18th century. Galois' father sided with the Revolution; he was actually driven to suicide by persecution from Royalists and priests. Young Evariste was a revolutionary firebrand. In his own lifetime, his profound and transformative contribution to algebra was neither accepted nor understood. He was an outsider to the mathematical establishment, which was then dominated by the great and famous Augustin-Louis Cauchy (1789–1857), a most pious, reverent Royalist and Catholic.

In the same era, Niels Abel (1802–1829) was also outstandingly brilliant, both in the field of polynomial equations and in the blossoming field of analytic function theory. He had the misfortune to be Norwegian, and therefore a foreigner and outsider to the mathematical grandees in Berlin and Paris. After great efforts to obtain some recognition, a professorship was finally offered to him. But by the time the offer arrived, he had died of tuberculosis exacerbated by overwork and poverty.

Bourbaki

One very influential group that started outside the mathematical establishment was Bourbaki. In 1934, a group of young mathematicians met for lunch in the Latin Quarter in Paris, at the cafe "A Capoulade," at 63 boulevard Saint-Michel, on the corner with rue Soufflot. (The café has since been replaced by an American fast-food outlet.) The group started as a project by

André Weil and Claude Chevalley to rewrite the obsolete analysis text of Edouard Jean-Baptiste Goursat (1858–1936) (We recount the early career of André Weil in chapter 3.) Chevalley was a devotee of avant-garde art and a member of an anarchist grouplet. They had a concrete project in mind: "to define for 25 years the syllabus for the certificate in differential and integral calculus by writing, collectively, a treatise on analysis."[5] Early additions to the group were Jean Dieudonné, Henri Cartan, and Jean Delsarte. Dieudonné became the secretary of the group. They whimsically decided to give their group a name, "Nicolas Bourbaki," a fictitious person who would become a famous "mathematician."

For decades every word published by Bourbaki received its final editing from the pen of Dieudonné. He was notoriously big, loud, and dogmatic. He not only remembered every word, he even remembered on what page every word had appeared. Another member, Henri Cartan, would become one of the major mathematicians of his time, specializing in Lie groups, functions of several complex variables, and coauthoring with Sammy Eilenberg the first major exposition of category theory. His father was the famous geometer Élie Cartan. Élie's father, Henri's grandfather, was a blacksmith.

Their objective soon became much more ambitious: to renew and modernize French mathematics, which they viewed as stuck in the classical French tradition focused on analytic function theory. Jacques Dieudonné wrote:

> The First World War was a dreadful hecatomb of young French scientists. When we open the war-time directory of the École Normale, we find enormous gaps which signify that two-thirds of the ranks were mowed down by the war. This situation had unfortunate repercussions for French mathematics. We others, too young to have been in direct contact with the war, but entering the University in the years after the war ended, should have had as our guides these young mathematicians, certain of whom we are sure would have had great futures. These were the young men who were brutally decimated and whose influence

was destroyed. Obviously, people of previous generations were left, great scholars whom we all honor and respect were living and still extremely active, but these mathematicians were nearly fifty years old, if not older. There was a generation between them and us. It is indubitable that a 50-year-old mathematician knows the mathematics he learned at 20 or 30, but has only notions, often rather vague, of the mathematics of his epoch, i.e., the period of time when he is 50. It is a fact we have to accept such as it is, we cannot do anything about it. [In 1970, when this article was published, Dieudonné was 64.] So we had excellent professors to teach us the mathematics of let us say up to 1900, but we did not know very much about the mathematics of 1920. The German mathematics school in the years following the war had a brilliance which was altogether exceptional . . . of whom we in France knew nothing. Not only this, but we also knew nothing of the rapidly developing Russian school, the brilliant Polish school, which had just been born, and many others. We knew neither the work of F. Riesz nor that of von Neumannn. The only exception was Élie Cartan; but being 20 years ahead of his time, he was understood by no one.[6]

Weil and Chevalley sought to emulate the spirit of the modern abstract algebra then being born at Göttingen under the leadership of Emmy Noether. The great textbook *Moderne Algebra* was written by the Dutch mathematician B. L. Van der Waerden with the participation of Emil Artin in the spirit of Emmy Noether. Van Der Waerden's book is a masterpiece of organization and conciseness. Everything is stated exactly when and where needed, nothing has to be repeated, and there's no need to refer to any other book. This style became the ideal of Bourbaki: complete rigor, complete self-containment, avoidance of unnecessary comments or explanations and any use of diagrams or illustrations, or use of geometric thinking. And rigorous avoidance of any contact with physics!

At the beginning they met monthly. They wanted to write collectively, and they aimed to introduce each topic with a general notion such as "field, operation, set, or group." Henri

Figure 6-1. Heinrich Behnke and Henri Cartan. Courtesy of Ludwig Danzer.

Cartan later wrote that the time appeared ripe for a comprehensive study of all important branches of mathematics, assuming nothing as given and making the basic interrelationships comprehensible, and so he and his friends decided to take this task upon themselves. He admitted that only the young could make such a bold decision. They were not unaware of the difficulties involved. Indeed, such an undertaking lay far beyond the powers of a single person. It would necessarily be a communal effort.

They did not opt for the usual division of labor, where each person writes on a single specialty. Instead, the entire group discussed each topic. This led to long debates, and in the end it was impossible to determine who had written what. The work became a collective effort in the true sense. The members brought in ideas and methods they had acquired abroad. They were teaching in provincial universities, which helped them to think independently of the centralized Parisian establishment. They consciously opposed existing institutions. They met outside the university, and they chose a publisher (Hermann) who at that time was marginal to the mathematicians who dominated the field.

Nevertheless, these men were an elite group in a highly hierarchical system, aspiring to the leadership of French mathematics. In fact, within 30 years Bourbaki became dominant in French mathematics, and very influential in many other countries. Their style and taste not only became standard for much advanced research, it even seeped down into college, secondary, and primary education. The "new math" project in the United States, known as the School Mathematics Study Group (SMSG), was one ramification of Bourbaki. In November 1959, at the Cercle Culturel de Royaumont, at Asnieres-sur-Oise, in France, at a conference on reforming French math education, Dieudonné rose to his feet with the cry, *"A bas Euclide! Mort aux triangles!"* ("Down with Euclid! Death to triangles!") Their idea was to throw geometry out of high schools and replace it with linear algebra. Dieudonné was by then over 50 and no longer an active member of Bourbaki.

The president of the Royaumont conference, and one of the guiding sprits of the new math, was Marshall Stone of the University of Chicago. Stone beautifully summarized his point of view by saying that a modern mathematician would characterize his subject as "the study of general abstract systems, each one of which is built of specified abstract elements and structured by

Figure 6-2. Jean Dieudonné of Bourbaki. Courtesy of Archives of the Mathematisches Forschungsinstitut Oberwolfach.

the presence of arbitrary but unambiguously specified relations among them." This abstractionist credo of Stone's is a perfect summary of Bourbaki'ism.

Bourbaki excluded any applied mathematics, and it totally ignored and avoided physics. Weil had lived in Göttingen in 1926, when the world of physicists there was bubbling with excitement as they gave birth to quantum mechanics. Yet, as he later wrote, he didn't notice what was going on around him at all.

The first volume in the Bourbaki series, *Elements of Mathematics*, came out in 1939, but then their meetings were disrupted by World War II. After the war, they published one or two volumes every year until 1983. Some volumes were quite long. The early volumes were about set theory, algebra, general topology, elementary calculus, and integration theory. At the beginning many members opposed including the topic of mathematical logic. Chevalley managed to change their minds. Later volumes treated Lie groups and commutative algebra. As Cartan described their working methods in the 1950s, members met three times a year at so-called Bourbaki congresses. Eight to

12 participants would gather in a quiet place far removed from the noise of the cities. Two meetings lasted a week, the third, during the summer vacation, lasted 14 days. They worked on average 7 or 8 hours a day; the rest of the time they walked and dined. It was not uncommon for all the participants to start talking at the same time.

The writing of these volumes could go through six or eight drafts, each draft written by a different author. Treatments of important topics were changed after sustained debates. Dieudonné wrote in 1970:

> Even when two men have a 20-year age difference, this does not stop the younger from hauling the elder, who he feels has understood nothing of the question, over the coals. One has to know how to take it, as one should, with a smile. . . . Certain foreigners, invited as spectators to Bourbaki meetings, always come out with the impression that it is a gathering of madmen. They could not imagine how these people shouting—sometimes three or four at the same time—about mathematics, could ever come up with something intelligent. . . . When we have seen the same chapter come back six, seven, eight or ten times, everybody is so sick of it that there is a unanimous vote to send it to press. . . . We are concerned then with replacing members affected by the age limit. . . . A youth of value who shows promise of a great future is quickly noticed. When this happens, he is invited to attend one of the Congresses as a guinea pig . . . the wretched young man is subjected to the ball of fire which constitutes a Bourbaki discussion. Not only must he understand, but he must also participate. If he is silent, he is simply not invited again.[7]

The deadly solemnity of their textbooks gives no hint of the carefree gaiety of their meetings with each other. When their semiregular publication of *Elements of Mathematics* became a commercial success, the royalties paid for travel expenses, wine, and extracurricular activities that enlivened the proceedings.

According to *La Tribu*, their internal newsletter, they played chess, table soccer, volleyball, or Frisbee. They went on mountain hikes, bicycle excursions, and swimming expeditions and even caroused in bumper cars. They went butterfly hunting or mushroom picking. They sunbathed, stuffed themselves with local delicacies, and drank until royally drunk—Armagnac, champagne, rum toddies, or wine. It was even reported that, once they had swallowed enough wine, members had been seen doing a virile French can-can or a lascivious belly dance.

Cartan emphasized that the close collaboration of Bourbaki required a feeling of community and friendship, complete candor, and good spirits, with each individual suspending his egotism for the good of the group. In a study of collaborative circles, the sociologist Michael Ferrell stresses the roles of thoughtful critics and enthusiastic audiences. Together with many others who have written about collaboration, he emphasizes that the most important feature is trust. The mathematicians who formed Bourbaki provided these resources for each other. They were patient listeners and powerful debaters, and they were remarkably able to sustain the vision that linked them.

The participants also thought carefully about how to keep group cohesion without sacrificing the freshness of new contributions. They established a rule that a member had to leave at age 50, so there was a succession of four generations of Bourbaki. The "alumni" remained identified with the group, even as they went on to high visibility as individuals in national and international mathematical establishments. The former outsiders became powerful insiders! The rise of the antiestablishment Bourbaki founders to become the establishment troubled some of them. But for years they stuck to the tasks ahead of them and ignored their internal tensions about the group's power and prestige.

Chevalley, the anarchist, was asked how it felt to have participated in a project that ultimately led to taking power. He answered that he felt a lot of resentment toward the members who produced this outcome. At the start, the congresses were paid for out of the members' pockets. But later, there were very substantial royalties. That was one cause of the degeneration. Then, before the war, it had been understood that one did not talk about career matters; "it was simply not done." But after the war, when they began to bring in young people, they naturally felt concerned about their careers. Little by little, they talked of everyone's career. The last straw, he said, was Dieudonné's propagandizing for mathematical reform.

As the group reached the fourth generation, the common focus and working style weakened, and members became more specialized in their interests. By the end of the 1970s, the style of Bourbaki had been so well propagated and understood that everyone knew how to write in this spirit. At this stage the group had run its course. Rather than initiating new works, the members decided to revise and update their previous volumes.

Alexandre Grothendieck, whose life we describe in chapter 4, was in the third generation of Bourbaki. In *Récoltes et Semailles* he wrote:

> It was certainly during the sixties that Bourbaki's "tone" slipped to increasingly prominent elitism, a change in which I surely took part. . . . I still remember my astonishment when I discovered in 1970 the extent to which Bourbaki was unpopular among the lower strata . . . of the mathematical world. The name had become more or less a synonym for elitism, for strict dogmatism, for a cult that favored "canonicalism" at the expense of living comprehension, for hermetism, for castrating antispontaneity, and that's not even all! . . . This group of exceptional quality exists no more. I don't know when it died, for it surely died without anyone noticing and sounding its death knell, not even within his heart of hearts. I suppose that an imperceptible degradation occurred in the members—everyone must have "got on in years," must have gone stale. They became important, prominent, powerful, feared, sought-after. Perhaps the spark remains, but innocence was lost along the way. . . . And respect was also lost along the way. When we had students, perhaps it was too

late for the best to be passed on—there was still the spark, but there was no longer innocence.[8]

In 1997 Pierre Cartier, a third-generation member, was asked why Bourbaki had published nothing new since 1983. He blamed a clash between Bourbaki and their publisher about royalties and translation rights that ended in a long, unpleasant lawsuit. But he added that the 1980s were a natural limit. After all, André Weil had insisted that every member should retire at 50, and so it made sense that Bourbaki also should retire when it reached 50. But the main reason, he thought, was that Bourbaki had achieved its stated goal: to provide foundations for all existing mathematics.

Cartier analyzed the ascent and decline of Bourbaki in terms of ideology; they resembled other strong ideologies of the 20th century whose leaders also believed in their limitless future.

> Bourbaki was to be the New Euclid, he would write a textbook for the next 2,000 years. . . . It is no accident that Bourbaki lasted from the beginning of the thirties to the eighties, while the Soviet system lasted from 1917 to 1989. The 20th century has been a century of ideology, the ideological age. When I began in mathematics the main task of a mathematician was to bring order and make a synthesis of existing material, to create what Thomas Kuhn called normal science. . . . Now we are again at the beginning of a new revolution. Mathematics is undergoing major changes. We don't know exactly where it will go, it is not yet time to make a synthesis of all these things—maybe in twenty or thirty years it will be time for a new Bourbaki. I consider myself very fortunate to have had two lives. A life of normal science and a life of scientific revolution.[9]

From a different perspective, one can compare Bourbaki's approach with that of logical positivism, a philosophy holding that metaphysical and subjective arguments not based on observable data are meaningless. Members in both of these thought communities fought for consistency, rigor, clarity, and intellectual boundaries. Their emphasis on pure rationality contrasted sharply with the chaotic world around them.

In analyzing scientific discoveries, biologist Ludvik Fleck described how division of labor, cooperation, preparatory work, technical assistance, mutual exchange of ideas, and controversy can produce a collective that contains much more knowledge than any one individual. But together with all of the advantages of a socially organized style of knowing, Fleck also describes how thought collectives can become rigid and resistant to new discoveries. In Bourbaki a similar growth and decline can be noted. They started as rebels against the established modes of thought in French mathematics in the early 20th century. As a carefully organized group, they were successful in developing a rigorous systemization of their field. But with new discoveries and the increasingly important interaction between physics and mathematics, their exclusionist approach lost its effectiveness. Contemporary mathematics is more multifaceted, it includes more varied theoretical and applied approaches. Bourbaki remains a fascinating example of a disciplined group effort to achieve intellectual transformation.

As our next example of a voluntary mathematical community, we describe the Anonymous Group, led by Paul Erdős in Budapest in the 1930s.

The Anonymous Group

The Hungarian mathematician, Paul Erdős, was legendary for his friendships and collaborations. Already as a young university student he met weekly in a park in Budapest with about 10 young people. Their meeting place was next to a statue of a 15th century historian whose name was not known; thus he and the Erdős group were called "Anonymous." One member, George Szekeres, who became a major Australian mathematician, remembered those meetings with great affection: "We met perhaps once a week and tried to go through the problems in a well-known book. These were collections of problems from mathematical analysis, and we tried to solve the problems, one after the other. It was a marvelous experience, I must say."

The group included several young women. One of them, Esther Klein, made an important contribution to the group when she brought to their attention a class of problems that later were known as "Ramsey theory." (At first they didn't know that such questions had previously been considered by Frank Ramsey in England.) In the simplest case of a Ramsey-type question, we consider a party with six guests. Then it's not too hard to prove that either there are three guests all of whom are already acquainted with each other, or else there are three guests no two of whom are previously acquainted. The numbers get much worse when you have more guests at the party. Szekeres was intrigued by the way in which Esther, his future wife, asked the question. He could solve the first part of the problem, after some struggle, but part of it is still unsolved. Erdős named this problem "the Happy Ending" because of Szekeres's and Klein's marriage. Szekeres and Klein fled from the Nazis to Australia, by way of Siberia and Shanghai. There they were able to inspire Hungarian-type problem contests.

The group members were all interested in problems of discrete mathematics—combinatorics, graph theory, and number theory. They first became acquainted with each other through the *Mathematics Journal for Secondary Schools*, which published challenging problems in each issue and listed the names and photographs of successful problem solvers.

As a result of the Great Depression and the quotas excluding Jewish academics, not one of these young mathematicians had a regular job. Erdős's parents were both gymnasium (high school) math teachers, and they could set up tutoring jobs for them. The members of the Anonymous Group included Pál Turán, Tibor Gallai, and George Szekeres, who would all become leading mathematicians in their own right—and Erdős's first collaborators. Other members were Márta Wachsberger, Géza Grünwald

(1910–1943), Anna Grünwald, András Vázsonyi, Annie Beke, Dénes Lazar, Esther (Eppie) Klein, and László Alpár.

Alpár went to France. There he was imprisoned as a Communist. When he was released at the end of World War II, he came home to Communist Hungary and was imprisoned again, under the Stalinist regime. After coming out of jail for the second time, he took up mathematics full time for the first time.

Turán served in a Fascist labor camp during World War II. Before and after that, he had a brilliant research career. At the time of his death in 1976, he had become a major figure in international mathematics. We quoted his account of doing mathematics in a labor camp in chapter 3.

Alpár, Erdős, Szekeres, and Klein had left Hungary before the Holocaust. Of those who had remained, only Vázsonyi, Gallai, and Turán survived.

Göttingen

Göttingen is far from the German capital Berlin; it is an idyllic little town on the slopes of the Hainberg. Its university was founded in 1737 by George II August, the Prince-Elector of Hannover and King of England, and so it's sometimes called "the Georgia Augusta." In 1866, when Hannover and its ally Austria were defeated in war by Prussia, Göttingen became part of Prussia. It was in Göttingen that the prince of mathematicians, Carl Friedrich Gauss, served for decades as director of the observatory. Gauss was followed by his devoted colleague Peter Gustave Lejeune Dirichlet. Georg Friedrich Bernhard Riemann, Gauss's student and Dirichlet's friend, was the third great mathematician who worked in Göttingen. But Gauss, Dirichlet, and Riemann had few pupils. They were not surrounded by a genuine mathematical community.

It was the wunderkind Felix Klein who conceived of creating a multidisciplinary scientific center at Göttingen. Klein had been a full professor at Erlangen at the unheard of early age of 23.

Students today learn about his Erlangen program, in which he unified and classified all geometries by their symmetry groups. In 1881 and 1882 Klein competed intensely with Henri Poincaré to develop the theory of automorphic functions. He wrote much later:

> The price I had to pay for my work was extraordinarily high—my health completely collapsed. In the next years I had to take long leaves and to renounce all productive activity. Things did not go well again until the autumn of 1884; but I have never regained my earlier level of productivity. I never returned to elaborate my earlier ideas. And later, when I was at Göttingen, I turned to extending the domain of my work and to general tasks of organizing our science. . . . My real productive activity in theoretical mathematics perished in 1882. All that has followed, insofar as it has not been purely expository, has been merely a matter of working out details.[10]

After he recovered from his breakdown, Klein was still a great teacher and expositor, but more than that, he became the *eminence grise*, the power broker and deal maker of German mathematics. Through a close friendship with Friedrich Althoff, the head of the Prussian system of higher education, from the 1880s to his death in the 1920s, Klein had a major say in who was appointed to what mathematics position in German universities.

Having in mind the growing needs of German industry and science, Klein imagined Göttingen as a different kind of mathematics center, one where mathematicians would welcome interaction with physics and engineering, even with biology and philosophy. It would provide an alternative to Berlin, where, under the control of the analyst Karl Weierstrass, the algebraist F. Georg Frobenius, and the number theorists Leopold Kronecker and Ernst Eduard Kummer, mathematics was pure.

David Hilbert epitomized Klein's notion of the kind of mathematician Germany needed. He joined the faculty at Göttingen in 1895. His personality, his scholarship, and his unusual breadth as a mathematician became the central core of one of the most outstanding mathematical communities in history. His first great contributions to mathematical research were in algebraic number theory. Then he turned to the foundations of geometry, to integral equations, to relativity theory, and to logic and the foundations of mathematics. In each area he made fundamental contributions that transformed the field. In each area he stimulated younger people to make their own important contributions. The full story of Göttingen is told in Constance Reid's books.

In his "Reminiscences from Hilbert's Göttingen," Richard Courant wrote:

If you read old chronicles, a Göttingen professor was a demi-god and very rank-conscious—the professor, and particularly the wife of the professor. Hilbert came to Göttingen and it was very, very upsetting. Some of the older Professors' wives met and said: "Have you heard about this new mathematician who has come? He is upsetting the whole situation here. I learned that the other night he was seen in some restaurant, playing billiards in the back room with Privat dozents." [The Privat dozent was lower than an "instructor" today, for he was paid nothing by the university; he received only whatever fees he could collect directly from his students.] It was considered completely impossible for a full professor to lower himself to be personally friendly with younger people. But Hilbert broke this tradition completely, and this was an enormous step toward creating scientific life; young students came to his house and had tea or dinner with him. Frau Hilbert gave big lavish dinner-parties for assistants, students etc. Hilbert went with his students, and also everybody else who wanted to come, for hour-long hikes in the woods during which mathematics, politics and economics were discussed.

One could also go and visit Hilbert in his garden. He spent his whole time gardening and in between gardening and little chores, he went to a long blackboard, maybe twenty feet long, covered so that also in the rain he could walk up and down, doing his mathematics in between digging some flower beds. All day one could observe him.

. . . He was a uniquely inspiring lecturer. . . . You had the chance to observe him struggling with sometimes very simple questions of mathematics, and finding his way out. This was more inspiring than a wonderfully perfect performance lecturing. The most impressive thing was the great variety, the wide spectrum of his interests. . . . He was a most concrete, intuitive mathematician who invented, and very consciously used, a principle: namely, if you want to solve a problem first strip the problem of everything that is not essential. Simplify it, specialize it as much as you can without sacrificing its core. Thus it becomes simple, as simple as it can be made, without losing any of its punch, and then you solve it. The generalization is a triviality which you don't have to pay too much attention to. This principle of Hilbert's proved extremely useful for him and also for others who learned it from him; unfortunately it has been forgotten.[11]

Hilbert was open to problems and ideas from everywhere, open to colleagues from other disciplines, and open to students from Hungary, America, Russia, and Japan. Hermann Minkowski and David Hilbert were fascinated by Einstein's theory of relativity, and Minkowski gave relativity theory its interpretation as a four-dimensional space-time manifold. Emmy Noether, the prime creator of modern abstract algebra, was fully accepted in Hilbert's research community, even as the academic bureaucracy denied her as a woman any of the recognition that would have been appropriate to her scientific stature. Grace Chisholm came from England. John Pierpont Morgan and later Willard Van Orman Quine and Saunders MacLane came from the United States. We will focus on three participants whose lives are not as widely known as those of Hilbert and Courant: Teiji Takagi, Fritz John, and Kurt Friedrichs.

One of the most interesting visitors to Göttingen was Teiji Takagi (1875–1960), a brilliant mathematics student in Tokyo.

In May 1898, Takagi received the following writ from the minister of education: "You are ordered to go to Germany in order to study mathematics for three years." Japan was still in an early stage of westernization in the late 19th century, and when Takagi was "ordered" to go study in Europe, it was a very high honor for a young Japanese scholar.

In the spring of 1900, after three semesters in Berlin, Takagi dropped in at Göttingen, where a friend was studying. Takagi was attracted by the work of Hilbert and changed his schedule to include a long stay in Göttingen. Takagi later wrote, "I was much astonished by the striking contrast in the atmospheres of the mathematics departments of Göttingen and of Berlin. In the former, once a week a meeting was held, and in attendance was a group of brilliant youths from all over the world, as if here were the center of the mathematical world."[12] This was unlike Berlin, which was more traditional.

In his book, *Theory of Algebraic Number Fields*, Hilbert had suggested that the "relatively Abelian fields" might be the most fascinating objects in this area, containing hidden, beautiful general laws. This book became the bible in Takagi's mathematical life. (Hilbert himself had already moved on, to study integral equations.) In Göttingen, Takagi made some progress on his project and published about it. He stayed in Germany for 5 years. Then he went home, married, became a very prestigious Japanese mathematics professor, and had six children. But he needed stimulation to be productive. In Japan there was no one to talk to about algebraic number theory. When World War I started, it became even worse, for now he no longer even received mathematical journals from Germany. Yet, strangely enough, this intensified isolation somehow did stimulate him! During the war, totally cut off from all research contact, he produced two great papers on "class field theory." Within a few years, his ideas came to be understood and appreciated in Europe. The key event was when Carl Ludwig Siegel (1896–1981), a German mathematician specializing in number theory, showed Takagi's work to his famous colleague Emil Artin.

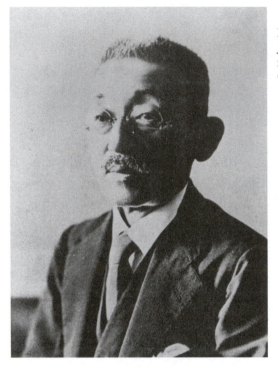

Figure 6-3. Taiji Takagi, Japanese algebraic number theorist. Courtesy of Ioan James.

In 1932 Takagi was sent to Europe as a delegate of the Science Council of Japan. In Vienna, Olga Taussky welcomed him and invited him to dinner at her apartment several times, providing him a warm reception with her mother and sister. In Hamburg, his student Shokichi Iyanaga introduced him to Emil Artin, and Takagi impressed Artin as being a modest but great scholar. Artin's wife, Natascha, later wrote, "I liked Takagi very much." In Göttingen, Takagi, accompanied by Emmy Noether, visited his master, Hilbert, who was struggling with liver disease. Takagi wrote, "Observing my old master grumbling as if speaking to himself, I wept in my heart." At the International Congress in Zurich, Takagi was one of the vice presidents. At this Congress, the award called the Fields Medal was established, with five judges including Takagi.

In the Hotel Eden by Lake Zurich, where Takagi was staying, he held a dinner party on September 11 or 12 which he later described as one of the best times in his long life. He wrote to his wife in Japan about this party, telling her how he had specially selected the wines to be served. The guests were Chevalley, Helmut Hasse, Shokichi Iyanaga, Y. Mimura and his wife, M. Moriya, M. Nagano, Emmy Noether, Olga Taussky, N. G. Chebotaryev, and B. L. van der Waerden. We can only speculate that Takagi's delight in assembling these famous mathematicians around one table may be related to his long periods of isolation from stimulating colleagues in his home country.

Two months later, the Nazis became the number one party in Germany. In 1934 Olga Taussky went to England and then in 1947 to the United States. Takagi survived the bombing and invasion of Japan. After World War II, the first letter to Takagi from abroad came from Olga, asking about his safety. Takagi passed away peacefully on February 28 in 1960, at the age of 84. About 1000 people attended his funeral.

One important visitor to Göttingen in the 1920s was the Russian topologist Pavel Sergeevich Aleksandrov (1896–1982), later head of the graduate math program during the "golden years" at Moscow State University. Aleksandrov wrote that one attractive feature of Courant's mathematical institute was the close relationships of its members, who were really woven into a single team. (In 1922 Courant founded the university's Mathematics Institute, although the building was not formally dedicated until 1929.) He further pointed out, although Courant's program was mainly directed towards mathematical physics, Emmy Noether's school of abstract algebra was far from applications. Nevertheless, both schools were closely linked by friendly connections between their members. These two schools, taken together, he thought, determined the face of Göttingen mathematics. He wrote of "their common enthusiasm, their unselfish love of mathematics, their awareness of its perfection as a striking creation of human thought and, following from this perfection, its inevitable unity. The idea of the uniqueness and intrinsic perfection of mathematics, and its unlimited cognitive force, which is necessarily directed towards the good of mankind, were Hilbert's credo as a scientist," said Aleksandrov, "and also the credo of the Courant and Noether schools."[13]

The great period of Göttingen was divided into two parts by World War I. The comfortable, prosperous times before the war became much harder in the years after 1918. At the university mess hall, soups and stews were ladled from a long trough, and students flirted with serving girls in the hope of getting larger portions. Constantly thinking about food made it hard to concentrate on their books. Yet the intellectual life was intense and idealistic, struggling for a firm position on the problems of the times—political, philosophical, religious, humanistic, artistic, and literary.

Two Göttingen students of the 1920s, Kurt Friedrichs and Fritz John, would become well-known, influential American mathematicians in the 1950s and 1960s. Kurt Friedrichs arrived at Göttingen as a young student in 1922. He was overwhelmed by the superior knowledge of the young lecturers and assistants at Göttingen, "such a bunch of people who knew everything about everything." He was captivated by the informal, exciting atmosphere around Courant. Even though Friedrichs was very retiring (not much at ease with himself or the rest of the world), Courant quickly recognized his gift. "I think it took keen observation as well as a really intense human interest on Courant's part to see what was there," said another of Courant's famous students.[14]

Even after Friedrichs had published several works and had several more awaiting publication, Courant admitted that Friedrichs still made a somewhat poor first impression. But in 1930 the Technische Hochschule in Braunschweig appointed Friedrichs to a full professorship.

Fritz John arrived at Göttingen in 1929. He wrote:

I arrived there as an almost pennyless student, with money scraped together by my hard working widowed mother. . . . I managed to survive due to the fact that some of the faculty extended a helping hand. . . . Practice sessions gave students an opportunity to come to the attention of the faculty. With Courant's help I obtained a stipend from the Studienstiftung des Deutschen Volkes, which relieved me of my main financial worries. Courant's lack of pomposity and his concern with students set an example to the faculty. Students whom he considered promising could count on his help, were invited to his house, and if suitable, participated in the musical activities of his family. He unselfishly strove to advance the cause of mathematics, though he undoubtedly enjoyed being in the center of things. . . . [Gustav] Herglotz gave beautifully polished lectures, seemingly on any subject under the sun, from celestial mechanics to geometry of numbers. In the course of time I attended many of his lectures and fell under his spell. . . . Sometimes his ethereal way of proving things in his lectures completely hid some simple direct access to the same result. One could only admire, but without hope of entering on one's own, this fantastic world of beauty, with its 19th century flavor. . . . Courant's course was the exact opposite. Courant's lectures lacked the glamour of those of Herglotz and of Hermann Weyl. They were, however, deeply stimulating, and offered a chance to participate in the creative process.[15]

These happy years were terminated by the rise of Hitler. Courant was Jewish. His heroic war record for Germany in World War I made no difference; he was kicked out of the institute he led in Göttingen. So were the great number theorist Edmund Landau and Felix Bernstein, the professor of statistics. Bernstein had once been vice-president of the local branch of the German Democratic Party, but he had given up politics when his support for the Weimar Republic hurt his standing in the academic community.

Among the many foreign visitors to Göttingen, several of the most famous were Jewish: the Swiss Paul Bernays, the Ukrainian Alexander Ostrowski, the Hungarians Theodor von Kármán and John von Neumann. Four German members—Richard

Courant, Ernst Hellinger, Max Born, and Otto Toeplitz—came from Breslau. Every one of these names became illustrious. Bernays became Hilbert's assistant and collaborator in mathematical logic. Von Kármán became one of the preeminent founders of modern aeronautics. Von Neumann, of course, was legendary, one of the true icons of modern mathematics. All were driven away.

But there had been a long history of antagonism in the university faculty between the liberal, cosmopolitan physicists and mathematicians and the nationalistic, reactionary faction that dominated the "humanist" faculties. The town itself was historically right-wing politically. A few faculty members like Bernstein had ventured into liberal or socialist politics, earning the lasting hatred of their opponents. By the end of the 1920s the student body was dominated by anti-Semitic and Nazi factions.

In Göttingen there was a large group of students around Hilbert who, as Courant wrote, really lived in complete dedication to the task of learning and studying. They were closely connected to each other and had much contact with the faculty. They spent a lot of time debating scientific and philosophical matters and trying to solve the mysteries of life. But the number of students increased, and gradually class distinction grew between those who had contact with assistants and the faculty and an anonymous mass of people who felt excluded. Courant thought that this certainly had something to do with the success of the Nazis. "The disenfranchised students who studied but didn't get anywhere saw others invited to dinner in the houses of the professors and included in swimming parties with the assistants and felt that they 'did not belong.' They formed, gradually, a large body of dissatisfied, sometimes quite intelligent elements. When the Nazis came, this was a wonderful reservoir for them. . . . In 1933 suddenly, to the great surprise of the faculty and older students, in many of the classes and in the seminars and at the university institutions, students turned up—but you didn't really know them—with the insignia of the Nazi party."[16]

When Courant was dismissed from his post as director of the Mathematics Institute, he was replaced by Helmut Hasse. A famous number theorist, Hasse was not a Nazi, merely a right-wing nationalist quite willing to serve the Nazis. He met suspicion and opposition from the student Nazi faction who wanted a real Nazi as director. At first the Hilberts spoke out against the new regime, and their friends remaining in Göttingen were frightened for their safety. But Hilbert and his wife didn't trust many of the people who were left, nor the new people who came, so after a while they too were silent.

The lives of Fritz John and Kurt Friedrichs were profoundly changed by the new policies of the German Reich. John's father was Jewish. His fiancée, Charlotte Woellmer, was not Jewish. Ten days after he received his Ph.D., he and Charlotte got married, not knowing whether it was still legal or safe to enter into a "mixed" marriage. They had to involve their families as little as possible, and there were anonymous denunciations against them. They felt trapped, with the door slowly closing. Finally, in the fall of 1933, Courant managed to secure a scholarship for John at Cambridge, and he emigrated to England safely in January. Charlotte followed 2 months later.

Friedrichs had no Jewish parent, but days after Hitler became chancellor he attended a ball in Brauschweig—the social event of the year—looking for a young Jewish girl he had noticed on the way home from his morning lecture. When he spotted her dark, straight hair, bobbed like that of the American movie actress Colleen Moore, he marched across the ballroom and invited Nellie Bruell to dance. Thus began the romance of his life.

In 1935 Friedrichs managed to visit Courant in New York. Courant promised to look for a job for him in the United States, no easy task at a time when many American Ph.D.'s couldn't find jobs. While Friedrichs was on his way home, the Nuremberg laws were passed, forbidding marriage between Aryans and non-Aryans. For a whole year, he and Nellie met infrequently and secretly. In 1936 he left Germany, allowed to take only 10 deutsch marks out of the country. To protect his parents, he told them nothing about his plans. Nellie had a French passport, and as soon as Friedrichs was safely out of Germany, she left for her father's home in Lyons. Friedrichs arrived in New York penniless. Courant found him a place to live, and Nellie came to join him. As soon as they could find a justice of the peace, they were married. Her outgoing, supportive nature and her talent for friendship perfectly complemented his needs.

After 1934 Hilbert stopped teaching but continued to work on logic. His friends and collaborators emigrated. His *Collected Works* appeared in 1935. He lived on in Göttingen, isolated from the international world of mathematics and suffering progressive loss of memory. In 1942 he fell and broke his arm. He died on February 14, 1943, of complications from the physical inactivity that resulted from the accident. Only a dozen people attended the funeral.

The Courant Institute

When Richard Courant, the leader and organizer of mathematics at Göttingen in the late 1920s and early 1930s was driven away, he went first to Cambridge and then to New York. There he established a graduate program at New York University (NYU) in the spirit of Göttingen, which during and after World War II became a leading world center of applied mathematics. One of the authors of this book, Reuben Hersh, is a graduate of that program. When I applied for admission as a graduate student in mathematics to New York University in the spring of 1957, I was interviewed by a soft-spoken man with a slight German accent. His name was Fritz John. After learning about my dubious qualifications, he suggested that I take advanced calculus during the summer. If I did all right, I could become a grad student in the fall. That fall, as a full-time graduate student in mathematics, I enrolled in Introduction to Applied Mathematics taught by Professor Cathleen Morawetz, who was the daughter of the well-known applied mathematician J. L. Synge.

Figure 6-4. Cathleen Morawetz (center), winner of the National Medal of Science, with colleagues. Courtesy of Sylvia Wiegand.

I had the good luck to have Fritz John as my professor in a course on complex variables. A couple of years later, while about halfway through my grad student career, one evening before going home from NYU's Mathematics Institute, I walked around the corner to La Maison Francaise, to hear a performance of Schubert's Trout Quintet. The room at La Maison Francaise was full. It felt like a salon, with an intense connection between listeners and performers. It was one of the most powerful musical experiences I have ever encountered.

The piano was played by Lenny Sarason, a fellow graduate student at the institute. Sarason had earned a master's degree under the composer Paul Hindemith before turning to mathematics. (Later we would share an office for a year as instructors at Stanford.) The violist was Lori Berkowitz, wife of one of my professors, Jerry Berkowitz, and daughter of the director of the institute, Richard Courant. The cellist was Jürgen Moser, Courant's son-in-law by way of Courant's other daughter. Moser would later become famous for contributions to celestial mechanics and dynamical systems.

The family atmosphere at the Mathematics Institute (renamed the Courant Institute when Courant retired a few years later) was not just a matter of chamber music. It also extended to a deep concern for the welfare of the students. Professor Bob Richtmyer's wife, Jane, was in charge of payroll. Once, when some foul-up delayed the teaching assistants' and research assistants' monthly checks, and I had the audacity to complain, Jane offered me a loan from her personal bank account. A needy graduate student I knew was helped by the institute, which hired his wife part-time for the electrical engineering group. Later, when it was rumored that he might leave for a better paying job, his pay was raised.

However, this warm family atmosphere was not enjoyed by all the students at the institute. There were insiders and outsiders. Many students were part-timers, taking one or two courses toward a master's degree while working at an electronics firm in New Jersey or Long Island. Such a student might feel left out when he noticed another student exchanging personal remarks with a professor or displaying a confident "at-homeness" in the lounge and the library. I was an outsider at first, and later as an insider I was still conscious of this status difference among the students.

I did editorial work for Richard Courant, the director, and I studied under his former students Fritz John and Kurt Friedrichs. I knew that Courant, Friedrichs, John, and also Lipman Bers were refugees from Nazi Germany. I even knew that Friedrichs and John had been Courant's students at Göttingen, a town somewhere in Germany, where the great, preeminent David Hilbert had reigned. But it was only much later, when writing this book, that I really appreciated what that meant. Friedrichs was elected to the National Academy of Sciences and in 1977 received the National Medal of Science, the United States' highest scientific award. After his retirement, the last talk Courant gave was at Friedrichs' 70th birthday. He spoke with great feeling about the man he had known for almost half

Figure 6-5. Kurt Otto Friedrichs of the Courant Institute. Courtesy of Archives of the Mathematisches Forschungsinstitut Oberwolfach.

a century as student, colleague, and friend: "He is one of the rare scientists whose intellectual and scientific development has never slackened but has gone forward continually. One of the very wonderful aspects of all this is the fact that even now at the age of seventy years, Friedrichs has not stopped or slackened in his endeavors and the radiant inspiration that emanates from him. . . . he has been and has become steadily more truly a great man of science. What a great human being he is everybody near to him knows very well."[17]

Fritz John died in New Rochelle, New York, on February 10, 1994. The following February the *Notices of the American Mathematical Society* published his obituary, written by Jürgen Moser, who had been both his colleague at NYU and his neighbor in New Rochelle. They often met at the train station in New Rochelle on the way to work at the institute. (Richard Courant and Lipman Bers also were long-time residents of New Rochelle.) During Moser's first year at NYU, John was working on one of his most important discoveries, the spaces of "bounded mean oscillation." Moser wrote that for him this

Figure 6-6. Fritz John and Jürgen Moser of the Courant Institute, in conversation. Courtesy of Archives of the Mathematisches Forschungsinstitut Oberwolfach.

discovery of John's was connected with an unforgettable personal experience. [While they were waiting for the train . . .] Fritz told Moser about his work on mappings with small strain and explained the subtle estimates for the derivatives it led to. Moser found it very interesting but did not appreciate the depth of this result at that time. But the next morning it just hit him that this inequality provided precisely the tool he needed for overcoming a major difficulty. "Never again," he wrote, "did I have such luck that a theorem was invented just at the time when I urgently needed it."

The obituary ends with a quote from Courant: "John is one of the most original and deep mathematical analysts of our time . . . completely uncorrupted by the activity of the marketplace, yet a full personality with wide intellectual interest."[18] Courant had already died in 1972.

Among U.S. math departments today, the Courant Institute of NYU remains special and different. Its faculty's interests range all the way from pure topics like topology and abstract algebra down to practical applications including meteorology, statistical mechanics, and mathematical physiology. And it still has an inclusive, welcoming attitude toward students and visitors and toward many different viewpoints in pure and applied mathematics. It preserves the Göttingen heritage, the tradition of Klein, Hilbert, and Courant. In response to a book review of the selected works of Peter Lax (my mentor at Courant), I received a message from a reader who had been reminded of his student days at Courant. He wrote that he clearly recalled Lax's friendly, approachable nature and the very high opinion all his students had of him as a person. He also had some interactions with his wife, Anneli Lax, and found her to be a "kind and gentle human being" too. He recalled how much Professor Friedrichs encouraged students to visit him in his office. The message ended, "He was inspirational. . . . Thanks for triggering my trip down memory lane."

The Golden Years of Moscow Mathematics

This is the title of a book published by the American Mathematical Society in 1992, edited by Smilka Zdravkovska and Peter L. Duren. It contains 12 articles by Russian mathematicians describing Moscow mathematics from the 1920s to the 1990s. The golden years were from about 1957 to 1968. In the preface, Zdravkovska writes that she was very fortunate to be an undergraduate student at the Mechanics and Mathematics Department (Mekh-Mat) of Moscow State University in the 1960s. It was an exciting environment where you learned as much from fellow students as from the professors. You could pick from among dozens and dozens of courses and seminars offered by first-rate mathematicians, and you also could teach and learn from the bright high school children in the mathematical *kruzhoks* (circles.) And of course the groups of students bound by close friendships shared other interests outside mathematics.

Vladimir Arnold, one of the most eminent living mathematicians, has written of the constellation of great mathematicians at the Mechanics and Mathematics Department. "It was really exceptional, and I have never seen anything like it at any other place." Professional mathematicians will recognize the names he lists. "Kolmogorov, Gelfand, Petrovsky, Pontryagin, P. Novikov, Markov, Gelfond, Lusternik, Khinchin, and P. S. Aleksandrov were teaching students like Manin, Sinai, Novikov, V. M. Alexeev, Anosov, A. A. Kirillov, and me. All these mathematicians were so different! It was almost impossible to understand Kolmogorov's lectures, but they were full of ideas and were really rewarding."[19] But in 1968 there came a marked chill in the atmosphere, as we will see below.

One of the most interesting contributions to this volume is by A. B. Sossinsky (Alyosha). He was born in Paris in 1937, into a family of Russian émigrés. On his father's side, he came from Russian nobility that can be traced back to the 16th century; however, the family had lost their land by the turn of the century.

His maternal grandmother, O. E. Kolbasina-Chernova, came from a well-off literary family (her father was a close friend of Ivan Turgenev). She became a Bolshevik, a "professional revolutionary." Sossinky's father, on the other hand, fought in the White cavalry against the Bolsheviks, and in World War II for the French Foreign Legion. He then became head of the Russian Verbatim Reports Section at the United Nations and settled in Great Neck, Long Island.

Sossinsky became fascinated with mathematics when the family still lived in France. He was 13 when the French curriculum introduced algebra and geometry. Geometry was his favorite subject, and he began "research" at age 14: "I 'proved' that Euclidean geometry is contradictory and 'showed' that the Universe is 'closed' in the sense that straight lines 'don't have two ends' but are 'like very big circles.' I was too shy to communicate my 'results' to my teacher (or to other grown-ups) but wrote them up in great detail, in a calligraphic handwriting, and sealed them in an envelope, meant to be opened to the world at large later on, when I would be old enough to be taken seriously."[20]

In 1954 he entered Washington Square College of NYU. After a disappointing freshman year, he planned to continue his education in Europe, either in Moscow or in Paris. In the summer of 1955 his family visited Russia. Their 2-month trip was quite a shock. They saw what the standard of living there was truly like and heard first-hand about the tragedy of Stalin's camps. The family returned to New York, but then in 1957, after another family summer in Moscow, Alyosha decided to stay there. It was a tough decision for him; he knew he could not expect to leave Russia again in the foreseeable future. His parents were neither supportive nor opposed. He did have naïve hopes that Khrushchev would soon be replaced by a younger, better educated, more liberal man, that some form of socialism with a human face would prevail.

During Sossinsky's undergraduate and postgraduate years (1957–1964), mathematics and mathematicians at Mekh-Mat flourished in a highly stimulating environment. The person most responsible for this was the rector of Moscow University, I. G. Petrovsky, an outstanding mathematician who headed the chair of differential equations for nearly two decades. Petrovsky is remembered even more for his honesty, his personal courage, and his remarkable ability as an administrator. He managed to concentrate a great deal of power in his own hands ("He has more clout than many Central Committee members, although he's not even in the Party," a well-informed administrator once told Sossinsky) and used it to expand and enrich the university in general, but especially the Mechanics and Mathematics Department, the apple of his eye.

The graduate math program was headed by the distinguished topologist P. S. Aleksandrov, who was always fond of and helpful to talented math students. With the powerful help of I. G.

Figure 6-7. I. G. Petrovsky, rector of Moscow
University and leading researcher in partial
differential equations. Courtesy of the Independent
University of Moscow and Moscow Center for
Continuous Mathematical Education.

Petrovsky, he was often successful in continuous struggles with
party bosses and the rank and file, especially when N. V. Efimov
became the Mekh-Mat dean (1959–1969). In fact, Efimov did
most of the infighting with the Party people in his friendly low-
key style. He was an able administrator, a very popular and
careful man.

Sossinsky adds the name of Andrei Kolmogorov, who held
no administrative position but was one of the world's outstand-

ing mathematical thinkers. "He symbolized the total scientific involvement, the intellectual probity viewed by many of us as the ideal for a mathematician." Sossinsky writes:

> It must be difficult for Western mathematicians to understand how, in a totalitarian society, scientific achievement as the main criterion for success in scientific institutions is something absolutely unusual. The usual criterion at the time in Russia was politics or ideology, not scientific truth. . . . Mekh-Mat, until the end of 1968, was a unique place, an oasis, a haven where the objective value of one's research work was one's best asset. This was understood and accepted by most students and teachers; it was an essential feature of the atmosphere at Mekh-Mat at the time. Our love of mathematics was, for most of us, part of a common outlook, characterized by anti-establishment political views and by a great interest in the artistic and literary life of the times and in active sports (especially mountain hiking, camping, canoe, cross-country and downhill skiing.)[21]

This use of mathematics as an escape from an oppressive reality is an example of a theme we developed in chapter 3.

In the same book, D. B. Fuchs, who was also a student at Mekh-Mat during its golden years, writes that the Brezhnev era began with several political trials, and the general atmosphere in the country was dreadful. "One of the important features of our political life in the late 1960s was 'signing letters.' After the trial of the writers Andrei Sinyavsky and Yuli Daniel, some groups of people wrote collective letters to various power institutions with various kinds of protest (from rather mild to very strong). Of course, the authors of the letters were punished, but the letters kept on being written. . . . In order to stop the campaign, the authorities had to choose one of the letters for an exemplary punishment of its authors. And it seems that they chose the letter about Esenin-Volpin."[22]

Alexander Sergeyevich Esenin-Volpin was the son of the great Russian poet Sergeii Esenin, a good mathematical logician, and

Figure 6-8. Andrei Kolmogorov (left), great Russian mathematician, with younger colleagues. Courtesy of Archives of the Mathematisches Forschungsinstitut Oberwolfach.

a confirmed dissident. In January 1968 he was taken to a *psik-hushka*—a special psychiatric hospital for political deviants.

> A letter in his defense was signed by 99 mathematicians. . . . The signers began to be pursued. There were meetings at various places where they were vilified by their colleagues. . . . Nobody knew what might happen. And we did a wise thing. A small group of mathematicians (including Shafarevich, Arnold, Tyurina, myself and some others) went to a remote skiing place in the Caucasus. We had no connection with the outer world and did not want to know what was going on in Moscow. And when we returned everything had been settled. There had been discussions in some high spheres, and it was decided to act without extreme measures. Thank God, nobody was arrested. Two people lost their primary jobs, and some people lost their secondary jobs. Many people had difficulties with promotions. No one was allowed to go abroad.[23]

We return to Sossinsky. "The year 1968 [the year of the Essenin-Volpin letter] was the turning point of many lives, including my own. It was the year of the May barricades in Paris, of draft-card burning and riots on American campuses, of the Prague spring crushed by Russian tanks. For me it was the year that put an end to my hopes and illusions, the year of dramatic events that marked the end of the Mekh-Mat golden era."

The letter was almost immediately published in the West, against the wishes of its authors and cosigners. The letter was the pretext for a crackdown at the Moscow University math department: the administration at Mekh-Mat and the party leaders were all subsequently replaced by hard-liners. Along with this crackdown came the organization of systematic anti-Semitic practices at the Mekh-Mat examinations.

Sossinsky remembers a boy named Kogan (a Russian version of Cohen) who was taking the entrance exam for the second time. He scored 5 (the maximum) on written math, 5 on oral math (surviving 4 hours of olympiad-level questions), and 5 on oral physics (where the two examiners were also out to get him.) That left the Russian literature essay, where even a passing grade would get him in. There were no spelling, grammar or stylistic mistakes in his essay, and Kogan had been an A student in literature, but he was given a 2 (= F) for "not clarifying the topic."

Sossinsky wrote, "I have never seen Kogan since. (Thank God. What would I have said to him?) For the first time I asked myself the question: what moral right did I have, as teacher, to be, if not an accomplice to, then a passive observer of, such practices?"[24]

In 1971 the Mekh-Mat party bureau decided to forbid Sossinsky to teach at School No. 12. In 1974 he wrote his resignation from the department. Then the chairman, P. S. Aleksandrov, invited him to his apartment to discuss the situation. Aleksandrov began with the following remarkable opening gambit: "Alyosha, traditionally we intellectuals of the Russian nobility have always placed our duty to the fatherland above our personal interests and feelings. A Russian nobleman does not leave a sinking ship—he fights to keep it afloat. It is people like Kolmogorov, like you and me, who have made this department into the unique scientific oasis that you know. Even in the Stalin years, we have always done all we could, swallowing our pride if need be. . . ."[25]

Sossinsky had known that Aleksandrov's parents were small-landed gentry, but he had expected anything but an appeal to values that 50 years of Bolshevik rule were supposed to have eradicated, especially in a careful and successful establishment scientist.

Sossinsky provides a psychological interpretation of how members of these outstanding institutions coped with these political pressures. He analyzed how the Mekh-Mat he knew and loved had been destroyed "by humiliating a student or a professor, forcing him to dig potatoes out of the mud by hand (as "voluntary" help for a local collective farm), making him hypocritically repeat, in public, obvious political lies about the system, the system succeeds in making this person lose his sense of self-respect. Then he becomes manageable."[26]

Talented people, who tend to be unpredictable and more difficult to control, are flunked at the entrance exams, or not recommended for graduate work, or not given positions in the department, unless their sense of self-respect is broken and they can prove their docility. "What the hard-line administrators wanted were good, competent, solid, stolid, servile mathematicians. And that's what they now have. There are very few world-class mathematicians holding a full-time position at Mekh-Mat today [1991], while there were dozens and dozens in 1968."[27]

The Jewish People's University

There is a little-known postscript to this story of Moscow mathematics. In 1978, a few years before the publication of the book *The Golden Years of Moscow Mathematics*, from which we have been quoting, a spontaneous, half-underground school of math called the Jewish People's University was created and served those excluded from Mekh-Mat. Bella Abramovna Subbotovskaya was a Jewish woman mathematician who in the mid-1950s had been a Mekh-Mat student. Now she organized classes covering the material taught to beginning students at Mekh-Mat—complex analysis, real analysis, topology, and algebra. With the help of her friends Valery Senderov (a well-known dissident) and Boris Kanevsky, she recruited first-class lecturers such as Dmitry Fuchs and Victor Ginzburg. She had a policy strictly forbidding any politics. Mathematics was taught, by volunteers, outside the official Soviet school system—that was all. About 350 students attended the school from 1978 to 1983. In March 1982, the famous U.S. topologist John Milnor taught there while visiting Moscow.

Andrei Zelevinsky, who is now at Northeastern University, writes that Bella Abramovna drew up lists of students, arranged places for class meetings, informed everyone of changes in scheduling, made sure that classes met and adjourned on time, brought chalk, and even made delicious sandwiches. "She accomplished all these tasks with a smile and without obvious efforts. Her mere presence created a wonderfully pleasant, warm, and homely environment. She took care of all practical everyday problems of all the instructors. By the way, it goes without saying that no one received any money for their work."[28]

Of course, such a thing could not go on without attracting the attention of the KGB. In 1982 Subbotovskaya was called in for questioning. It is believed that she refused to cooperate in testifying against Kanevsky or Senderov. Then a strange thing happened. At about eleven o'clock, on the night of September 23, 1982, as Bella was walking home down a quiet street after visiting her mother, a truck ran into her at high speed and then drove away. A few minutes later an automobile stopped nearby, and soon after that an ambulance appeared and took her dead body to the morgue. The funeral service was attended by a few friends and family members. No one dared to voice

Figure 6-9. Bella Abramovna Subbotovskaya. Courtesy of Ilya Muchnik.

their suspicions. Kanevsky and Senderov were imprisoned, for 3 and 5 years, respectively. The Jewish People's University ceased to exist.

Association for Women in Mathematics

Women mathematicians have faced different challenges. They confront centuries-old discrimination based on their gender. The founding of the Association for Women in Mathematics (AWM) was motivated by women's determination to be accepted as equals in mathematical circles and by the difficulty they have had in obtaining good academic jobs. At the 1971 Joint Mathematics Meetings women activists called for a caucus. They were aware of the very discouraging statistics about their status in their field as reflected by the scarcity of women in the program. None of the invited, hour-long speakers were women, and only 5 percent of the short talks were given by females. A similar gender disparity existed in academic jobs; only 1 of 10 promotions listed that year in *Notices of the American Mathematical Society* was given to a woman. At the instructors' level they were better represented: 33 percent of the jobs listed were held by women. The American Mathematical Society (AMS) was run by men, with no major elected positions held by their female peers.

Soon after the caucus was established, the organizers transformed it into an independent organization called the Association for Women in Mathematics. Mary Gray, a professor at the American University, was instrumental in its formation and became its first president. As part of her leadership, she wrote the first issue of the organization's newsletter, a publication that provides the vital link that builds this community to this day.

For Lenore Blum, a difficult early experience at Berkeley transformed her "from being naive about the politics for women in academia, to taking an active part in helping women navigate the difficult terrain of such a traditionally male field as mathematics."[29] It was at that time that Blum became active in local and national organizations, and she formed strong ties with other women in science, including Judy Roitman. She served from 1975 to 1978 as the third president of the Association for Women in Mathematics. Her search for a supportive community led her to combine political activism and mathematical research. She believed that rather than adjusting to existing established organizations, one needed to create new ones.

The AWM's members' activities are wide-ranging: they document the status of women in mathematics; they influence public policy and that of granting agencies; their speakers' bureau maintains close connections with high schools and colleges; and they address the many forms of bias held against women interested in mathematics.

The organization also celebrates the lives and contributions of women mathematicians (for instance, by a colloquium

Figure 6-10. Mary Gray (left), founder of the Association for Women in Mathematics, with friends. Courtesy of AWM, AWM Newsletter 23(3), 25.

honoring Emmy Noether) and dedicates prizes and offers scholarships in acknowledgment of their most famous predecessors. AWM's political and educational impact is impressive for an organization with 4000 members. Government and scientific groups frequently consult its members. The organization also helps develop the mathematical potential of high school and college women through conferences and workshops.

Many of these kinds of activities are regularly performed by effective professional organizations in science. But AWM goes beyond these usual contributions; it is a community of practice and shared beliefs, which is emotionally significant in the lives of its members. Jean E. Taylor and Sylvia Wiegand described AWM as a "passionate organization." The readers of the bimonthly newsletter comment that "each issue 'recharges' them and helps them fight feelings of isolation." The persistence of discrimination contributes to these feelings. In a special issue of *Notices* it was reported that while female enrollment in advanced math classes has increased since the establishment of AWM, increase in tenured appointments at prestigious universities is still proceeding slowly.[30]

It is not surprising that this discrimination has psychological consequences. In the same issue of *Notices*, D. J. Lewis writes

Figure 6-11. Lenore Blum (far right) and colleagues. Courtesy of the American Mathematical Society.

that there are clear indications that at every level, from middle school to doctoral programs, women generally are less confident in their mathematical abilities than men. Successful women do receive encouragement and assurance of their abilities from parents and instructors, yet for many successful women in mathematics, there is always a doubt that they are as good as they are. "Perhaps, some of this self-doubt arises because the general public has come to view mathematics as masculine and early on women perceive themselves as outsiders to the mathematical world."[31]

Many women still feel isolated and embattled in their roles as mathematicians. Not only do they need to prove that they are effective in their chosen profession, and prove it again and again, they also need to show that they have not abandoned their sense of identity as women. At AWM meetings, in their publications, and through their friendships with other members, they are able to discuss these dilemmas. They share their concerns about "the two-body problem"—that is, the professional challenge, when married to a male mathematician, of trying to find jobs together.

In biographies, personal accounts, and presentations of male mathematicians, reference to their personal lives is limited. This is not the case in writings by and about women. Throughout the book *Complexities*, written by members of the AWM, emotional conflicts and issues are freely and frequently discussed. At the same time, the volume includes technically demanding papers in mathematics, and facts and figures about the slowly changing status of women in the profession. In this way, the contributors overcome the usual dichotomy between intellect and emotion.

Women mathematicians are subject to discouraging cultural messages. One is the belief that they lack an aptitude for mathematics. Another is the argument that math lacks the intuitive sensibility so important to women who cherish relationships. In confronting these opinions, women in the AWM have asserted that mathematics provides them with a rewarding profession, a way to use their intellect maximally, without jeopardizing a full emotional life and broad humanistic interests.

Other contributors to *Complexities* write of the impact of math and computer camps for high school students, of special women's mentoring programs at prestigious institutions, and of networking and support groups including the AWM. The difference that such activities have made in the lives of women aspiring to be mathematicians is made clear by Judy Roitman when she recalls the past: "It was not uncommon for major women mathematicians to be unemployed; young women were routinely discouraged; the few who persevered were usually treated badly; and role models were few and far between."[32]

There have been important changes throughout the 20th century regarding women's access to mathematical careers. The most important of these is a change in university policies. At the beginning of the century, women in many countries could not take classes for official credit. They could only audit classes, and only with permission of the lecturer. And even when they were allowed to audit, women often were not ready for the university courses because of the inferior education they had received earlier.

Many universities continued to discourage women applicants in the decades before World War II. Even after such policies became illegal, attitudes toward women interested in mathematics remained prejudicial. The ensuing psychological injury has healed slowly. The sense of inadequacy, from which many women continue to suffer, manifests itself in a more fragile self-concept. The studies of Benbow and Stanley further contributed to societal stereotypes. But since then, meta-analyses of large databases show a steady shrinking of gender differences in mathematical achievement. Most research on the cognitive abilities of males and females, from birth to maturity, does not support the claim that men have greater intrinsic aptitude for mathematics and science. These findings do not receive the publicity that greeted earlier studies focusing on gender differences. But they contribute to make equity a reality, through intervention programs for women and minorities. Most important are the encouragement provided by caring mentors and the group interaction of students drawn from previously marginalized groups.

At the beginning of this chapter we asked, What kinds of communities do mathematicians form? The range of groups in which mathematicians function professionally is very broad. There are some communities that enrich their members' lives in very significant ways. Whether within institutions or formed intentionally outside traditional universities, they provide meaningful connections for their members. Some have a clear, shared vision. Others provide enthusiasm and cooperation. Their existence and importance refute the myth of the solitary, isolated nature of mathematical life.

Bibliography

Abelson, P. (1965). Relation of group activity to creativity in science. *Daedalus* (summer), p. 607.

Albers, D. J., & Alexanderson, G. L. (Eds.) (1985). *Mathematical people: Profiles and interviews*. Boston: Birkhäuser.

Aleksandrov, P. S. (2000). "A few words on A. N. Kolmogorov," in *Kolmogorov in Perspective*. Providence, R.I.: American Mathematical Society, pp. 141–144.

Beaulieu, L. (1993). A Parisian café and ten proto-Bourbaki meetings (1934–1935). *Mathematical Intelligencer 15*(1), 27–35.

Beaulieu, L. (1999). Bourbaki's art of memory. *Osiris 2*, 2nd series, vol. 14, *Commemorative Practices in Science: Historical Perspectives on the Politics of Collective Memory*, pp. 219–251.

Bell, E. T. (1937). *Men of mathematics*. New York: Simon & Schuster.

Birman, J. (1991). In her own words. *Notices of the American Mathematical Society 38*(7), 702.

Blum, L. (2005). AWM's first twenty years: The presidents' perspectives. In B. A. Case & A. M. Leggett (Eds.). *Complexities: Women in mathematics*. Princeton N.J.: Princeton University Press, pp. 80–97.

Bourbaki, N. (1948). Foundations of mathematics for the working mathematician. *Journal of Symbolic Logic 14*, 1–14.

Bourbaki, N. (1950). The architecture of mathematics. *American Mathematical Monthly 57*, 221–232.

Brown-Walker, J. (2005). A double dose of discrimination. In B. A. Case & A. M. Leggett (Eds.). *Complexities: Women in mathematics*. Princeton, N.J.: Princeton University Press, pp. 189–190.

Cartan, H. M. (1980). Nicolas Bourbaki and contemporary mathematics. *Mathematical Intelligencer 2*(4), 175–187.

Case, B. A., & Leggett, A. M. (Eds.) (2005). *Complexities: Women in mathematics*. Princeton, N.J.: Princeton University Press.

Courant, R. (1980). Reminiscences from Hilbert's Göttingen. *Mathematical Intelligencer, 3*(3), 154–164.

Dieudonné, J. A. (1970). The work of Nicholas Bourbaki. *American Mathematical Monthly 77*, 134–145.

Fleck, L. (1979). *Genesis and development of a scientific fact*. Chicago: University of Chicago Press.

Fuchs, D. B. (1993). On Soviet mathematics of the 1950s and 1960s. In S. Zdravkovska & P. L. Duren (Eds.). *Golden years of Moscow mathematics, History of Mathematics,* vol. 6. Providence, R.I.: American Mathematical Society, 220–222.

Gowers, T., & Nielsen, M. (2009). Massively collaborative mathematics. *Nature 461*, 879–881.

Grothendieck, A. (1986). *Récoltes et Semailles*. Unpublished manuscript.

Guedj, D. (1985). Nicholas Bourbaki, the collective mathematician: An interview with Claude Chevalley. *Mathematical Intelligencer* 7(2), 18–22.

Heims, S. J. (1980). *John von Neumann and Norbert Wiener: From mathematics to the technologies of life and death*. Cambridge, Mass.: MIT Press.

Helson, R. (2005). Personal communication.

Henrion, C. (1997). *Women in mathematics*. Bloomington, Ind.: Indiana University Press.

Hersh, R., & John-Steiner, V. (1993). A visit to Hungarian mathematics. *Mathematical Intelligencer* 15(2), 13–26.

Honda, K. (1975). Teiji Takagi: A biography. Commentary. *Mathematica Universitatis Sancti Pauli* XXIV-2, 141–167.

John, F. (1992). Memories of student days in Göttingen. *Miscellanea Mathematica*. New York: Springer, pp. 213–220.

Klein, F. (1979). Development of mathematics in the 19th century. Translated by M. Ackerman. In R. Hermann. *Lie groups, history, frontiers and applications*, vol. IX. Brookline, Mass.: Math Science Press.

Kline, M. (1974). *Why Johnny can't add*. New York: Random House.

Lewis, D.J. (1991). Mathematics and women: The undergraduate school and pipeline. *Notices of the American Mathematical Society* 38(7), 721–723.

Lui, S. H. (1997). An Interview with Vladimir Arnol'd. *Notices of the American Mathematical Society* 42(2), 432–438.

MacTutor web Site. Mathematical biographies: Élie Joseph Cartan, Henri Paul Cartan, Pierre Émile Jean Cartier, Claude Chevalley, Jean Alexandre Eugene Dieudonné.

Mashaal, M. (2006). *Bourbaki: A secret society of mathematicians*. Providence, R.I.: American Mathematical Society.

Mathias, A. R. D. (1992). The ignorance of Bourbaki. *Mathematical Intelligencer* 14(3), 4–13.

Moser, J. (1995). Obituary for Fritz John, 1910–1994. *Notices of the American Mathematical Society* 42(2), 256–257.

Pier, J. P. (Ed.) (2000). *Development of mathematics 1950–2000.* Boston: Birkhäuser.

Reid, C. (1983). K. O. Friedrichs 1901–1982. *Mathematical Intelligencer* 5(3), 23–30.

Reid, C. (1992). *Hilbert.* New York: Springer-Verlag.

Reid, C. (1996). *Julia: A life in mathematics.* Washington, D.C.: MAA Spectrum.

Roberts, S. (2006). *King of infinite space.* New York: Walker and Company.

Roitman, J. (2005). In B. A. Case and A. M. Leggett (Eds.). *Complexities: Women in mathematics.* Princeton N.J.: Princeton University Press, p. 251.

Rota, G. C. (1987). The lost café. *Los Alamos Science*, Special Issue, p. 26.

Rowe, D. E. (1986). "Jewish Mathematics" at Göttingen in the era of Felix Klein. *Isis* 77, 422–449.

Rowe, D. E. (1989). Klein, Hilbert, and the Göttingen mathematical tradition. In K. M. Oleska (Ed.). *Science in Germany: The intersection of institutional and intellectual issues. Osiris* 5, 189–213.

Rowe, D. E. (2000). Episodes in the Berlin-Göttingen rivalry, 1870–1930. *Mathematical Intelligencer* 22(1), 60–69.

Schechter, B. (1998). *My brain is open: The mathematical journeys of Paul Erdős.* New York: Simon and Schuster.

Segal. S. L. (2003). *Mathematicians under the Nazis.* Princeton, N.J.: Princeton University Press.

Senechal, M. (1998). The continuing silence of Bourbaki—An interview with Pierre Cartier, June 18, 1997. *Mathematical Intelligencer* 20(1), 22–28.

Shifman, M. (Ed.) (2005). *You failed your math test, Comrade Einstein.* Singapore: World Scientific.

Socha, K. (2005). Mathematics: Mortals and morals. In B. A. Case & A. M. Leggett (Eds.). *Complexities: Women in mathematics.* Princeton N.J.: Princeton University Press, pp. 393–395.

Sossinsky, A. B. (1993). In the other direction. In S. Zdravkovska & P. L. Duren (Eds.). *Golden years of Moscow mathematics, History of mathematics*, vol. 6. Providence, R.I.: American Mathematical Society, pp. 223–243.

Szanton, A. (1992). *The recollections of Eugene P. Wigner as told to Andrew Szanton*. New York: Plenum Press.

Szpiro, G. G. (2007). Bella Abramovna Subbotovskaya and the "Jewish People's University." *Notices of the American Mathematical Society 54*(10), 1326–1330.

Tikhomirov, V. M. (1991). *Moscow mathematics 1950–1975*. Basel: Birkhäuser.

The Scientist (1992). Academy criticism of a foreign associate stirs debate over NAS role and policies. 6(19), 1–18.

Ulam, S. (1976). *Adventures of a mathematician*. New York: Scribner.

van der Waerden, B. L. (1937). *Moderne algebra*. Berlin: Springer.

Weil, A. (1992). *The apprenticeship of a mathematician*. Basel: Birkhäuser.

Zelevinsky, A. (2005). Remembering Bella Abramovna. In M. Shifman (Ed.). *You failed your math test, Comrade Einstein*. Singapore: World Scientific.

Zdravkovska, S., & Duren, P. L. (Eds.) (1993). *Golden years of Moscow mathematics, History of mathematics*, vol. 6. Providence, R.I.: American Mathematical Society.

+ 7 +

Gender and Age in Mathematics

In Chapter 5 we quoted Hardy's dictum, "Mathematics is a young man's game," which has become a catch phrase. In this chapter we consider both aspects of that catch phrase: "young" and "man." What happens to a mathematician as he or she gets older? Do women have the same career opportunities as men? Do they follow the same rhythms and patterns in their work as their male colleagues? In the second half of the chapter, we consider age and aging of mathematicians. In the first half, we survey the past and present status of women mathematicians.

Women in Mathematics

By the expression "a young man's game" Hardy did not mean to exclude women. The famous British analyst Mary Cartwright was his student. In 1941, when Hardy wrote his *Apology*, normal usage was to write "man" either in the sense of "masculine" or in the sense of "human." Nowadays, of course, we say "young person" if we mean to include both sexes. Hardy was not only a pacifist and an atheist but can even be counted as an early feminist, as attested by his active support for the English-American mathematician-turned-biophysicist Dorothy Wrinch.[1]

Here we will give a brief survey of women in mathematics, in the past and today. We will see that until recent decades it was indeed overwhelmingly a "man's game," because women were excluded. Not primarily by mathematicians—but first, by parents' demands to conform to social expectations, and then by exclusionary policies of university administrations. Full equality or equity is still a goal to be struggled for.

We begin with short accounts of the lives of three great female mathematicians who overcame tremendous obstacles in order to become mathematicians: Sophie Germain, Sofia Kovalevskaya, and Emmy Noether.

Marie-Sophie Germain (1776–1831)

In chapter 1 we told how at the age of 13 in revolutionary Paris Germain became fascinated with Archimedes and with mathematics, despite the fierce opposition of her parents. After they realized they could not defeat Sophie, her father funded her research and supported her efforts to break into the community of mathematicians. For many years this was the only encouragement she received. Like many other women who practiced mathematics in later times, she devoted herself solely to her profession and never married.

In 1794 the École Polytechnique opened in Paris, reserved for men. Sophie assumed the identity of a former student, Antoine-August Le Blanc. Although M. Le Blanc had left, the Academy continued to print lecture notes and problems for him. Sophie obtained Le Blanc's notes and problems and submitted solutions under Le Blanc's name to the supervisor of the course, Joseph-Louis Lagrange. But Lagrange noticed that M. Le Blanc's solutions showed remarkable improvement, and Germain was forced to reveal her identity. Lagrange was astonished and became her mentor and friend.

Germain wrote to Legendre about problems suggested by his 1798 *Essai sur le Théorie des Nombres* and the subsequent Legendre-Germain correspondence became virtually a collaboration. Legendre included some of her discoveries in a supplement to the second edition of the book. Several of her letters were later published in her *Oeuvres Philosophiques de Sophie Germain*.

However, Germain's most famous correspondence was with Carl Friedrich Gauss (1777–1855). She had developed a thorough understanding of the methods presented in his 1801 *Disquisitiones Arithmeticae*. Between 1804 and 1809 she wrote a dozen letters to him, initially adopting again the pseudonym "M. LeBlanc" for fear of being ignored because she was a woman. During their correspondence, Gauss gave her number theory proofs high praise, an evaluation he repeated in letters to his colleagues. Germain's true identity was revealed to Gauss only after the 1806 French occupation of his hometown of Braunschweig. Recalling Archimedes' fate and fearing for Gauss' safety, she contacted a French commander who was a friend of her family. When Gauss learned that the intervention was due to Germain, who was also "M. LeBlanc," he wrote to her with delight: "When a person of the sex which, according to our customs and prejudices, must encounter infinitely more difficulties than men to familiarize herself with these thorny researches, succeeds nevertheless in surmounting these obstacles and penetrating the most obscure parts of them, then without doubt she must have the noblest courage, quite extraordinary talents and superior genius."[2]

She became interested in Fermat's Last Theorem and adopted a new approach to the problem. She considered those prime numbers p such that $2p + 1$ is also prime. (Germain's primes include 5, because $11 = 2 \times 5 + 1$ is also prime, but not 13, because $27 = 2 \times 13 + 1$ is not prime.) For values of n equal to these Germain primes, she proved that if $x^n + y^n = z^n$, then x, y, or z must be a multiple of n. In 1825 Johann Peter Gustav Lejeune Dirichlet and Adrien-Marie Legendre independently proved that the case $n = 5$ has no solutions. They based their proofs on Sophie Germain's work. This remained the most important result on Fermat's Last Theorem from 1738 until the work of Kummer in 1840.

After Gauss turned from number theory to applied mathematics, Germain also stopped working on number theory and instead took up a major challenge in the theory of elasticity. In 1808 the German physicist Ernst F. Chladni had visited Paris where he exhibited the so-called Chladni figures that are produced by a layer of sand on top of a vibrating plate. The Institut de France set up a prize competition with the following challenge: "Formulate a mathematical theory of elastic surfaces and indicate just how it agrees with empirical evidence."[3] Lagrange said that the mathematical methods available were inadequate to solve it, but Germain nevertheless spent the next decade attempting to derive a theory of elasticity, competing and collaborating with some of the most eminent mathematicians and physicists. She first submitted a manuscript in 1811, in which she originated the concept of mean curvature. Even though she was the only entrant in 1811, she did not win the award. Lagrange was one of the judges, and he corrected errors in Germain's calculations and came up with an equation that he believed might describe Chladni's patterns. The deadline was extended by 2 years, and again Germain submitted the only entry. She showed that Lagrange's equation did yield Chladni's patterns in several cases, but she could not give a satisfactory derivation of Lagrange's equation from physical principles. For this work she received an honorable mention. Germain's third attempt in the reopened contest of 1815 finally won the prize, a medal of 1 kilogram of gold. As a result, she became the first female mathematician to attend sessions at the French Academy of Sciences.

To public disappointment, she did not appear at the award ceremony, perhaps because Poisson, her chief rival on the subject of elasticity, was a judge of the contest. He had sent a laconic and formal acknowledgment of her work but avoided any serious discussion with her and ignored her in public. Later, when others built upon her work, and elasticity became an important scientific topic, she was closed out.

She was stricken with breast cancer in 1829 but, undeterred by that and the fighting of the 1830 revolution, she completed papers on number theory and on the curvature of surfaces (1831). In a eulogy her friend Count Libri-Carducci, wrote of her "unfailing benevolence, which caused her always to think of others before herself . . . in science, never thinking of the advantages that success procures. She rejoiced even when she saw her ideas made fruitful on occasion by other persons who adopted them."[4]

Sofia Vasilyevna Kovalevskaya (1850–1891)

We have already met Kovalevskaya twice: first, in chapter 1, describing her childhood fascination with mathematics, stimulated by the wallpaper in her nursery; and then in chapter 5, describing her friendship with her mentor Karl Weierstrass. Now we will fill in the missing parts of her biography.

When Sofia started studying mathematics under a tutor, Y. I. Malevich, she "began to feel an attraction for my mathematics so intense that I started to neglect my other studies."[5]

So her father ordered a stop to the math lessons. But she found a copy of Bourdeu's *Algebra* and read it at night when her parents were asleep. A year later, a neighbor, a Professor Tyrtov, gave her family a physics textbook he had written. When Sofia didn't understand the trigonometric formulas, she tried to decipher them, and rediscovered the way in which the notion of sine was historically developed. So Professor Tyrtov tried to convince Sofia's father to let her study mathematics. But he refused for several more years.

When Sofia reached the age of 18, she wanted to go to a university. But Russian universities didn't admit women. In fact, Russian women weren't even allowed to live apart from their families without written permission from a father or husband! Her father had provided her with tutoring, including calculus at age 15, but he would not give her permission to go abroad.

So she escaped by making a "marriage of convenience" with an amenable young student of paleontology, Vladimir Kovalevskii. They left Russia with her sister, Anyuta. For the next 15 years, frequent quarrels and misunderstandings between husband and wife resulted in exasperation, tension, and sorrow.

Sofia went to Heidelberg to study mathematics and natural science, while her sister went to Paris. But the University of Heidelberg did not admit women! Eventually Sofia convinced the authorities to let her attend lectures unofficially if she got permission from each lecturer. Sofia immediately attracted attention with her uncommon mathematical ability. After 2 years she went to Berlin to study with Karl Weierstrass. She had to study privately with him, as the university in Berlin would not allow women even to attend class sessions.

By the spring of 1874, Kovalevskaya had written three papers any of which Weierstrass deemed worthy of a doctorate, one each on partial differential equations, Abelian integrals, and Saturn's rings. The first of these was published in *Crelle's Journal* in 1875.[6] In 1874 the University of Göttingen awarded Kovalevskaya a doctorate, summa cum laude without examination and without her having attended any classes at that university. Sofia Kovalevskaya was the first woman in Europe to earn a doctorate in mathematics. Her doctoral dissertation is today called the Cauchy-Kovalevskaya theorem.

She returned to Russia and settled in St. Petersburg, her husband's hometown. Despite the doctorate and strong letters of recommendation from Weierstrass,[7] Kovalevskaya was unable to obtain an academic position. Her best job offer was to teach arithmetic to elementary school girls, and she remarked bitterly, "I was unfortunately weak in the multiplication table."[8] For the next 6 years, she devoted herself to her family (her daughter Sofia Vladimirovna, nicknamed Fufa, was born in 1878), to scientific journalism, and to promoting women's right to higher education. She also wrote fiction, including a novella, *Vera Barantzova*, which was translated into several languages. (Sofia's interest in literature went back to her childhood, when

Dostoyevsky was a regular guest in her family's home.) In 1890 she wrote in a letter, "It seems to me that the poet must see what others do not see, must see more deeply than other people. And the mathematician must do the same."[9]

In St. Petersburg she was visited by a fellow student of Weierstrass, the Swede Gosta Mittag-Leffler. He later wrote, "When she speaks her face lights up with such an expression of feminine kindness and highest intelligence that it is simply dazzling. Her manner is simple and natural, without the slightest trace of pedantry or pretension. She is in all respects a complete 'woman of the world.' As a scholar she is characterized by her unusual clarity and precision of expression. . . . I fully understand why Weierstrass considers her the most gifted of his students."[10]

She came back to mathematics in 1880, when Chebyshev and Mittag-Leffler invited her to speak at an international congress in St. Petersburg. In that year she returned to Berlin and visited Weierstrass, and took up the study of light propagation in anisotropic media. In 1882 she wrote three articles on the refraction of light.

In the spring of 1883, Vladimir committed suicide. The couple had been separated for 2 years. To escape from guilt, Kovalevskaya immersed herself in mathematics.

In 1883 Mittag-Leffler overcame opposition and obtained a position for her as *privat docent* at the University of Stockholm. In June 1889, after surviving an extremely hostile atmosphere, she was awarded a lifetime professorship, the first woman to hold a chair at a modern European university.

At Stockholm she taught the latest topics in analysis, became an editor of the new journal *Acta Mathematica*, took responsibility for liaison with mathematicians in Paris and Berlin, and helped to organize international conferences. She again wrote reminiscences and dramas, as she had when she was young. In 1886 the French Academy of Sciences announced the topic of the *Prix Bordin*: significant contributions to the motion of a rigid body. Kovalevskaya entered and won the prize with a paper, *Mémoire sur un cas particulier du problème de le rotation d'un corps pesant autour d'un point fixe, où l'intégration s'effectue à l'aide des fonctions ultraelliptiques du temps.* In recognition of the brilliance of this paper, the award was increased from 3000 to 5000 francs. In 1889 her further research on this topic won a prize from the Swedish Academy of Sciences, and she was elected a corresponding member of the Russian Imperial Academy of Sciences. The Russian government had repeatedly refused her a university position, but the rules at the Imperial Academy were changed to allow a woman to be elected. Two years later, early in 1891, Kovalevskaya died of influenza and pneumonia, at the height of her mathematical powers and reputation.

Emmy Amalie Noether (1882–1935)

Next we turn to Emmy Amalie Noether, who persisted in the face of tremendous obstacles to become one of the greatest algebraists of the 20th century. Emmy was the eldest of four children. Her father, Max, was a distinguished mathematician and a professor at Erlangen. Her mother, Ida Kaufmann, came from a wealthy Cologne family. Her mother taught her to cook and clean. She was sent to the Höhere Töchter Schule in Erlangen, a kind of young ladies finishing school, where she studied German, English, French, and arithmetic. She took piano lessons and loved dancing at parties with the children of her father's university colleagues. After high school and further language study, she was certified to teach English and French in girl's schools in Bavaria.

But Emmy Noether never became a language teacher. At the age of 18, she decided to take classes in mathematics at the University of Erlangen, where her father was a professor and her brother, Fritz, was a student. She could not be a regular student. In 1898 the academic senate of the University of Erlangen had resolved that admitting female students would "overthrow all academic order." But Emmy was permitted to audit classes. After 2 years she went to the University of Göttingen and attended lectures by Blumenthal, Hilbert, Klein, and Minkowski. Then in 1904, without ever having been a regular undergraduate student, she passed the qualifying exam to matriculate in mathematics at Erlangen as a doctoral student. In 1908 she received a doctorate summa cum laude for a dissertation entitled "On Complete Systems of Invariants for Ternary Biquadratic Forms," written under the direction of her father's colleague Paul Albert Gordan, whom she had known since childhood. Gordan's lifework was calculating explicit algebraic formulas for the invariants. Following this constructive approach of Gordan, Noether's thesis listed systems of 331 covariant forms. Later she scornfully rejected this dissertation as *Formelsgestrupp*—a jungle of formulas.

She could not teach at Erlangen because they did not hire women professors. But she helped her father, teaching his classes when he was sick. Because of his disabilities, he was grateful for his daughter's help. And soon she started publishing papers on her own work. Ernst Fischer (known for the Riesz-Fischer theorem) had succeeded Gordan in 1911 and influenced Noether away from Gordan's calculational approach and toward Hilbert's abstract approach. Noether's reputation grew quickly. In 1908 she was elected to the Circolo Matematico di Palermo, then in 1909 to the Deutsche Mathematiker-Vereinigung.

In 1915 Hilbert and Klein invited Noether back to Göttingen. Although her specialty was algebra, not mathematical physics, Emmy's first accomplishment in Göttingen was a central result in theoretical physics, now known as Noether's theorem. It states a correspondence between differentiable symmetries and conservation laws and led to new formulations for several concepts in Einstein's general theory of relativity. The American theoretical physicist Lee Smolin recently wrote, "The connection between symmetries and conservation laws is one of the great discoveries of twentieth century physics. But I think very few non-experts will have heard either of it or its maker—Emily Noether, a great German mathematician. It is as essential to twentieth century physics as famous ideas like the impossibility of exceeding the speed of light."[11]

Hilbert and Klein convinced her to remain at Göttingen. In order to allow Noether to lecture, Hilbert announced her courses under his name. For example, in the catalogue for the winter semester of 1916–1917, there appears: "Mathematical Physics Seminar: Professor Hilbert, with the assistance of Dr E Noether, Mondays from 4–6, no tuition."

Hilbert and Klein had a long fight to have her officially included in the faculty. It was war time, and her opponents asked, "What would the country's soldiers think when they returned home and were expected to learn at the feet of a woman?" Hilbert answered, "I do not see that the sex of the candidate is an argument against her admission as a *Privat dozent*. After all, the university senate is not a bathhouse."[12]

Emmy was a pacifist and hated the war. In 1918 the Kaiser surrendered and Germany became a republic. Then Noether and all women received the vote, and a year later, at the age of 37, she finally became a *privat dozent*—an unpaid faculty member, entitled to try to collect fees from her students. Three years later she was given a higher title, "unofficial associate professor," but still she remained unpaid, and as long as she remained at Göttingen, this never changed. There were several reasons for this. Not only was she a woman, she was also a Jew, a Social Democrat, and a pacifist. However, she did act as thesis adviser for a number of Göttingen Ph.D. students.

After 1919 Noether developed a new point of view in algebra: focus on simple, general, axiomatic properties shared by many algebraic structures. In this way she produced a new theory of "ideals" that helped turn the algebraic specialty called ring theory into a major mathematical topic. In *Idealtheorie in Ringbereichen* (1921) she proved the fundamental result: in any commutative ring with an ascending chain condition, the ideals can be decomposed into intersections of primary ideals. (In chapter 4, in our discussion of Grothendieck, we explained the meaning of "ideal" in a ring.)

Beyond its importance as a basic theorem in algebra, this work gradually changed the way mathematicians think. Noether's conceptual approach to algebra led to a body of principles unifying algebra, geometry, linear algebra, topology, and logic. "She taught us to think in simple, and thus general, terms . . . homomorphic image, the group or ring with operators, the ideal . . . and not in complicated algebraic calculations," said her Russian coworker P. S. Aleksandrov.

After spending the year 1924 studying with Noether, the Dutch mathematician B. L. van der Waerden wrote his very influential book *Moderne Algebra*. Most of the second volume consists of Noether's work. In 1926 André Weil visited Göttingen. In subsequent decades, his influential Bourbaki group adopted Noether's axiomatic style as the correct mode for all of pure mathematics.

From 1927 on, Noether collaborated with Helmut Hasse and Richard Brauer on noncommutative algebras. Along with Emil Artin and Helmut Hasse, she founded the theory of central simple algebras.

Her keen mind and infectious enthusiasm made Emmy Noether an effective teacher for those students who could keep up with her. The ones who caught on to her fast style became loyal followers known as "Noether's boys." Much of her work appeared in papers written by colleagues and students rather than under her own name.

What was Emmy like as a person? "Warm, like a loaf of bread," wrote Hermann Weyl. "There radiated from her a broad, comforting, vital warmth. She was fat, rough and loud, but so kind that all who knew her loved her." Her students were like her family, and she was always willing to listen to their problems.

In 1933 the Nazis came into power in Germany and demanded that all Jews be thrown out of the universities. In April 1933 she was denied permission to teach. Hermann Weyl wrote, "Her heart knew no malice; she did not believe in evil—indeed it never entered her mind that it could play a role among men. This was never more forcefully apparent to me than in the last stormy summer, that of 1933, which we spent together at Göttingen . . . her courage, her frankness, her unconcern about her own fate, her conciliatory spirit, were in the midst of all the hatred and meanness, despair and sorrow, surrounding us, a real solace."[13]

Noether's brother, Fritz, was fired from his position teaching mathematics at the Technical University in Breslau, on complaints that his presence contradicted the Aryan principle. He was offered a teaching position at Tomsk in Siberia and moved there with his family in 1934. In 1937 he was arrested as a German spy and sentenced to 25 years in prison. While in prison, he was accused of "anti-Soviet propaganda," and he was executed at Orel on September 10, 1941. Over 40 years later, after the death of Stalin, his son was informed of his father's complete "rehabilitation."[14]

Friends tried to get Emmy a position at the University of Moscow. Then came an offer of a professorship at Bryn Mawr College. There was a difficulty, because Noether had no interest in undergraduate teaching and Bryn Mawr had no prospect of creating a permanent position without undergraduate teaching. (Their graduate program in mathematics had only four faculty and five students.) But the Rockefeller Foundation and the Emergency Committee to Aid Displaced German Scholars paid her first year's salary, and her supporters, including Birkhoff, Lefschetz, and Wiener, succeeded in persuading the college to extend her appointment.

Emmy took four of the students under her wing and taught them in a mixture of German and English. She also lectured at the Institute for Advanced Study in Princeton. The grant was renewed for two more years in 1934. This was the first time that Emmy Noether was ever paid a full professor's salary and accepted as a full faculty member. For the first time, she had colleagues who were women. Anna Pell Wheeler, the head of the mathematics department at Bryn Mawr, became a great friend of hers. Emmy told people that these were the happiest years of her life.

In 1935 she developed a uterine tumor and scheduled surgery during a college break at Bryn Mawr. She died during or shortly after the surgery from sudden and unexplained complications. Her death surprised nearly everyone, as she had told only her closest friends of her illness. Noether had never married, and she had no relatives in the United States. Her ashes are buried in the cloisters of Thomas Great Hall on the Bryn Mawr campus. A coed gymnasium in Erlangen specializing in math has been named after her.

Contemporary Women Mathematicians

What is it like today to be a woman in mathematics? Certainly things have changed a lot since the times of Germain, Kovalevskaya, and Noether. At the beginning of the 20th century, women in many countries could not take university classes for official credit. They could audit only with permission of the lecturer. Some women experienced skepticism or even opposition from their teachers when expressing interest in mathematics during their early school years.

Many universities continued to discourage women applicants in the decades before World War II. Even after such policies became illegal, attitudes toward women interested in mathematics remained prejudicial. Pregnant women, and women with small children, found it hard to continue their studies. It took the help of committed teachers to encourage female students to persist. The ensuing psychological injury has healed slowly.

Since World War II, the number of women mathematicians has increased. As women started to share their experiences through meetings, caucuses, and the establishment of the Association for Women in Mathematics, they were able to develop a diversity of solutions to the challenge of balancing family and work lives.

For younger women today, whose careers started in more recent decades, the path has been less difficult. The goal of equal opportunity and active effort for equal representation in all academic fields has made mathematical life more welcoming for women. In the United States, "affirmative action" actually requires all university departments, including mathematics, to actively seek females for faculty positions. Among math students, women are now respectably represented.

Women and minorities are entering graduate programs in increasing numbers and are making major contributions. In the last two decades, there has been a steady growth in female participation in mathematics, in both undergraduate and graduate programs. One-third of Ph.D.s in mathematics are now granted to women, nearly doubling the numbers since the eighties.[15]

Today there are scores of women in tenured positions in university mathematics departments, and many of them, more than one can count on both hands, are internationally recognized. While sexist attitudes linger, women are assuming leadership positions in larger numbers in national and departmental organizations.

Critical to the successful outcome of each woman's story are supportive colleagues in the mathematics community. Margaret Murray found that nearly all the women she interviewed had had at least one college teacher who had encouraged and influenced them, even in the days when women in mathematics challenged the prevalent social norms.

But many, probably most of these women, had to resist and overcome prejudice and discrimination on their way to success. And even with equal opportunity and affirmative action in place, the number of women in leading positions in mathematics remains small, much smaller than in other fields of science like biology. Among faculty, they are still a small minority. In the most prestigious departments, the crème de la crème of the Ivy League, there are practically none.

In her excellent book *Women in Mathematics*, Claudia Henrion asks, "First, why is it that women continue to be significantly underrepresented in mathematics, particularly at the highest levels? And second, why is it that even the most successful women in mathematics, those who have already made it by standard measures of success, often continue to feel (to varying degrees) like outsiders in the mathematics community?"[16] In the course of considering graduate programs herself, Henrion writes, she was advised to look elsewhere by the chair of a mathematics department that had a large, prestigious program: "You're too normal; you wouldn't fit in here."[17] Most of the women Henrion interviewed reported variations on that experience at various times in their careers.

In the popular mind, mathematics is still a male occupation. Fewer girls than boys choose to take advanced math courses or to major in math. And even among those women who do graduate work in math and receive doctorates, there is still a lower rate of promotion and research productivity. As we saw earlier in writing about the AWM in chapter 6, there are three distinct major issues that keep down the number of female mathematicians. First, women have greater difficulty in establishing the connections and support systems that are essential for scientific success; second, the desire and need to have children forces women to interrupt their research careers in ways that are hard to compensate for; and third, the lingering stereotypes and social expectations of gender create self-doubt and hinder the active fighting for success and recognition that successful male mathematicians often display.

It often takes intense effort to create support groups for women, while men, it seems, are largely free to get on with doing mathematics. Female graduate students of mathematics report difficulty in connecting to the study circles and conversation groups of their male fellow students. If there are no other females in her particular mathematical specialty, a woman in graduate school needs to have exceptional tact or perseverance to be accepted into male social circles.

As Henrion says in her first chapter, "It is so much a part of the normal course of events for men to be tied into the mathematical community that it is easy for them to take it for granted, not even recognizing the way in which community is invisibly operating behind the scenes: in helping them to get jobs, in brainstorming sessions with colleagues about ideas, in the exchange of news about recent theorems in bars or bathrooms, in the suggestion of each other's names as speakers or journal editors."[18]

One of the most famous of the female mathematicians interviewed by Henrion is Karen Uhlenbeck, a professor at the University of Texas in Austin and a winner of the prestigious "genius" MacArthur award. As a girl, Uhlenbeck was an extreme tomboy, who thought she might become a forest ranger as an adult. Her mathematical style has always been utterly individualistic, finding surprising connections between ideas in theoretical physics and abstract geometry or nonlinear partial differential equations. Her career path through graduate school and faculty positions seems to have developed without much conscious planning on her part, but rather by other people noticing her work and being impressed by it. Even with a MacArthur fellowship and membership in the National Academy of Sciences to her credit, Uhlenbeck says "[I was] not able to transform myself completely into the model of the successful mathematician because at some point it seemed like it was so hopeless that I just resigned myself to being on the outside looking in."[19]

A second issue is the question of family life. A century ago, the few women who aspired to academia expected to remain unmarried. Neither Germain, Kovalevskaya, nor Noether had the standard husband-and-children family life. Today, that limitation is not accepted by most female mathematicians. But having children is almost bound to disrupt a mother's professional activities. Even pregnancy can already be problematic. Childbirth and child care in early infancy usually cause a sharp break in one's study or one's research unless one is lucky enough to have an exceptionally supportive husband, mother, or mother-in-law. The myth of the young, energetic mathematician contributes to the pressure young women (and men) feel, despite the fact that there are many examples of prominent mathematicians who did excellent work in their later years. If the focus were not so much on the young mathematician, it would be more effective to design programs with women in mind.

Figure 7-1. Karen Uhlenbeck, a MacArthur award recipient and leading researcher in mathematical physics. Courtesy of Dirk Ferus.

Different female mathematicians have given different answers to the question: Is it best to have children in your last year of grad school? (as did the logician Lenore Blum and the combinatorialist Fan Chung Graham). Or to have children after college and defer graduate study? (the path taken by the topologist Joan Birman of Barnard College–Columbia University, who received her Ph.D. at the age of 41). Or to have them much later, in your late thirties or even forties, after you have achieved tenure?

Chung decided to forego maternity leave during her second year at Bell Labs and had her second child during her 4-week vacation (during which she wrote a paper). Her manager wondered if she would quit because she was having a child. He was unaware that she already had a child who was 2 years old. It was crucial that she had full-time child care and a supportive husband, Ron Graham. When one of Lenore Blum's professors saw her with her 4-month old baby, he said "Where did that child come from? Whose is it?"[20] He had been oblivious to her pregnancy and giving birth.

Joan Birman is an outstanding researcher in topology, specializing in braids and knots. She earned her doctorate at the Courant Institute as a student of Wilhelm Magnus, after years working at engineering firms and raising three children. She had returned to graduate school at New York University later in life, starting part-time in a master's program. She points out that since many women have children during the very years

Figure 7-2. Fam Chung, American combinatorialist. Courtesy of Archives of the Mathematisches Forschungsinstitut Oberwolfach.

that traditionally are considered the most productive years for a mathematician, it would be more appropriate to look at their productivity over a longer time span, recognizing that women may need to enter the mathematical research pipeline later in life. Birman reported a relatively smooth path to a Ph.D. and eventually a position as a respected research mathematician. But many women find themselves having children at about the same time that they come up for tenure, "and the men are working like lunatics on their mathematics, and the women, they see this choice, and if you give it a little less effort, your research is dead."[21]

In the linear trajectory from graduate school through post-doctoral study, assistant professor, associate professor, and full

Figure 7-3. Joan Birman, topologist at Columbia University. Photograph by Joseph Birman.

professor, any deviation is suspect. Taking a couple of years off from mathematics makes it very difficult to return. There are very few reentry points. Some colleagues may still perceive you as being "not fully serious about your work" if you have a child. The silent message—"You're either a mathematician or a mother, you can't be both"—goes with the assumption that men do the math and women do the mothering. To a young woman who wants to become a mathematician, there may seem to be no favorable period for having children. The mathematical community needs to find ways of dealing with this conflict.

One outstanding success at reentering research after a long hiatus was Lenore Blum, who has recently collaborated with Steve Smale and Mike Shub on theoretical numerical analysis using real numbers rather than integers. Early in her career she accepted a job at a women's college in Berkeley, Mills College, after a promised position at the University of California, Berkeley, mysteriously failed to materialize. At Mills she developed and created an exemplary math program and became deeply involved in increasing female participation in mathematics. She helped organize and was president of the AWM. Then, after 13 years devoted to this work, at the age of 40, she returned to research and managed to start a highly successful research career.

The third often mentioned negative factor is psychological. Every mathematician is subject to self-doubt. Am I smart enough? Can I make it? Can I still do it? It seems that such self-doubt is more severe or more dominant for many women. The social stereotype that mathematics is a male domain is involuntarily internalized. It may take many years for a woman to feel at ease and accepted by her mathematical community.

Many young mathematicians grow up in environments where their interest and ability are not recognized. There is a sense of vulnerability from which many women continue to suffer. In *Complexities*, a volume by and about women mathematicians, Katherine Socha writes of such self-doubt and the difference caring mentors make: "I worried about being good enough and only after several years of graduate school did I develop the perspective and confidence to ask 'good enough by whose standards?'"[22] Socha comments, "It was only my stubborn determination to prove my intellectual mettle that kept me enrolled in mathematics classes."[23]

Judy Roitman said of herself, "I think I tend to be a little pushy and strident because I had to be in graduate school to be heard. So I am generally not a very good colleague. I don't think I socialize correctly or work well with other people. Basically, I'm too sensitive to not being taken seriously, and I tend to simply fade away and give up."[24]

The widely publicized studies of Benbow and Stanley (1980), reporting male superiority in tested performance, further contributed to gender stereotypes. But since then, meta-analyses of

Figure 7-4. Manuel, Lenore, and Avrim Blum. All three are now professors at Carnegie-Mellon University. Courtesy of Archives of the Mathematisches Forschungsinstitut Oberwolfach.

large databases show a steady shrinking of gender differences in mathematical achievement.[25]

Claudia Henrion conducted extensive interviews with half a dozen prominent female mathematicians. She found a remarkable range of different ways of achieving success within the male-dominated mathematical structure.

Mary Rudin, whom we quote extensively in chapter 8, spent most of her mathematical career as a self-styled "amateur," holding part-time jobs while devoting equal time and effort to running a household and raising four children. She was placed in her first jobs by her mentor, R. L. Moore, with no effort on her part. Her husband, Walter, was a well-known tenured professor specializing in a totally different field of mathematics.

In 1971, when it became embarrassing that one of their best known mathematicians was a female and a part-time instructor, the math department at Wisconsin in one single step elevated her to full professor with tenure, a promotion she appreciated but had neither asked for nor expected. She was 47 years old.

The sharpest contrast with Rudin's story is that of Vivienne Malone-Mayes, an African-American mathematician whom we have met in chapters 1 and 5, and will meet again in chapter 9 in connection with her denial of admission to the course taught by Robert Lee Moore. Mayes earned her bachelor's and master's degrees at Fisk, in Nashville, Tennessee. Fisk is a traditionally black institution (TBI) comparable to Howard University in Washington D.C. as one of the highly ranked, prestigious TBIs. At Fisk she was inspired by two teachers, Evelyn Boyd Granville, who was one of the only two African-American females who had earned a math Ph.D. at that time, and Lee Lorch, a white professor who was a legendary fighter for civil rights and equality for Black Americans. After graduating from Fisk, she taught at Paul Quinn College and at Bishop College, where, like her mentors Granville and Lorch, she sought out talented Black students and encouraged them to aspire to do advanced graduate work for a doctorate. As a consequence, she herself was challenged to go back to school for a Ph.D. She had grown up just a few blocks away from Baylor University, but when she applied for admission to their graduate program, she was told that Blacks were not admitted. In the early 1960s the University of Texas in Austin had been forced to give up its whites-only admission policy. Mayes applied and was admitted to the graduate program at Austin, a path that many in her community thought was absurd for a Black woman, and certainly not practical for getting a job. By the time she had completed her Ph.D. at Austin, Baylor had reversed its policy and offered her a faculty position! She taught at that overwhelmingly white institution from 1966 to 1994, and in 1971 the Baylor student congress elected her outstanding faculty member of the year.

However, Mayes' successful teaching career at Baylor came at the price of virtually complete isolation from the rest of the faculty. In fact, during her whole time there she was paid much less than comparable white instructors. When she asked about teaching summer classes, she was told by the department chair, "I haven't even offered those to the white male faculty yet."[26] Throughout her career Mayes was sustained by her religion and by her goal of serving her students and her African-American community.

In all these very different stories, three elements of success are present: having persistent self-confidence, connecting with mentors and collaborators, and being in the right place at the right time, when a situation opened up that permitted continued mathematical work and advancement.

What Can Be Done?

Contributors to *Complexities* write of the impact of math and computer camps for high school students, of special women's mentoring programs at prestigious institutions, and of networking and support groups including the AWM. Most important are the encouragement provided by caring mentors and the group interaction of students drawn from previously marginalized groups.

Joan Birman told Henrion, "I guess that if I could see any solution to the non-participation of women in mathematics, it would be, first of all, if women were able to think about going back to mathematics at a later point, and if there was a practical way for them to do this . . . and if the whole community was ready to accept this as another option—all of these things would help."[27] Joan Birman would not have been able to get a Ph.D. at Columbia, where she is now a professor, because she needed to start graduate school as a part-time student, and Columbia does not allow part-time graduate students in mathematics.

But at Smith College the Ada Comstock program allows older women who left school in order to raise a family to finish their bachelor's degree. Certain graduate programs, like the one at New York University, are receptive to older students or those who have taken some time off. The National Science Foundation has a program for women in mathematics who are returning to research.

There should be ways for mathematicians to have a part-time status during certain periods of their careers, perhaps in graduate school or as professors. This is one way of allowing people to have children and yet remain professionally active, even if it is at a reduced pace for a few years. For extenuating personal circumstances, such as having children, the tenure-track period could be lengthened. Many colleges and universities are beginning to institute such policies. Day care at mathematics meetings, flexible teaching schedules, and regular day care at colleges and universities are important. And of course, a change in attitude in the mathematics community. Attitudes can be more important than formal policies in determining whether women return to mathematics. The mathematics community must convey a clear message that having children is not in conflict with a career in mathematics.

Aging Mathematicians

We will report on a survey in which mathematicians told us about their experiences and feelings about aging, and we will present a critique by Louis Mordell of Hardy's views on the subject.

In chapter 2 we quoted Hardy's beautiful and melancholy essay, *A Mathematician's Apology*. At 60, says Hardy, he is too old to have new ideas because "mathematics is a young man's game." So he is reduced to writing books instead of doing a mathematician's proper business, discovering and creating new mathematics. First, as to "young." In this very same essay

Hardy says that his own best time for mathematical creation was in his forties, in collaboration with John Littlewood and Srinivasa Ramanujan. And the long productive life of his collaborator Litlewood is a striking refutation of Hardy's "young man" theory. Littlewood outlived Hardy by 30 years and was over 70 when with Mary Cartwright he wrote one of his most intricate and important papers, dealing with Van der Pol's equation and its generalizations. It is 110 pages of hard analysis. He called that paper "the Monster" and said: "It is very hard going and I should not have read it had I not written it myself." Years later it was recognized as an early breakthrough in the discovery of chaos. (See chapter 2.) Over a decade later, at age 84, he published, in the first issue of the *Journal of Applied Probability*, "very precise bounds for the probability in the tail of the binomial distribution. The bounds he gave are still the best."[28] Littlewood's last paper appeared in 1972, when he was 87.

Hardy's *Apology* was challenged by Hardy's successor at Cambridge, Louis Joel Mordell (1888–1972). Mordell's life is another counterexample to Hardy's pessimism. Mordell had arrived at Cambridge in 1906 as the wunderkind son of a poor Hebrew scholar in Philadelphia. He retired in 1953, at age 75, but almost half of Mordell's 270 publications appeared after his retirement! Mordell is reported to have said modestly, "I did work in my 70s many a younger man would have been proud to have done."[29] (Nowadays, Mordell is most often mentioned in connection with his famous conjecture in number theory.)[30] After he retired, Mordell visited universities around the world, reaching a total of 190. In 1971, well into his eighties and still traveling enthusiastically, he attended a number theory conference in Moscow, went on an Asian tour, and lectured in Leningrad. A few months later he died at home in Cambridge.

A year earlier, in October 1970, Mordell published his critique of Hardy's *Apology*. To begin with, he disputes Hardy's self-belittling dictum, "The function of a mathematician is to

Figure 7-5. Louis Joel Mordell and Gábor Szegö. Courtesy of Dolph Briscoe Center for American History, The University of Texas at Austin. Identifier: di_05555. Title: Mordeel and Szego. Date: 1958/08. Source: Halmos (Paul) Photograph Collection.

do something, to prove new theorems, to add to mathematics, and not to talk about what he and other mathematicians have done."[31]

Mordell goes on,

> The real function of a mathematician is the advancement of mathematics. Undoubtedly the production of new results is the most important thing he can do, but there are many other activities which he can initiate or participate in. Hardy had his full share of these. He took a leading part in the reform of the mathematical tripos some sixty years ago. Hardy was twice secretary and president of the London Mathematics Society. . . .We all know only too well that with advancing age we are no longer in our prime, and that our powers are dimmed and are not what they once were. Most of us, but not Hardy, accept the inevitable. We can perhaps find pleasure in thinking about some of our past work. We can still be of service to younger mathematicians.[32]

Mordell mentions Littlewood, Chapman, Besicovitch, and Davenport, all still active in their sixties, seventies, and eighties.

Mordell even turns Hardy's notion of mathematical beauty against him. Hardy's words have often been quoted: "A mathematician, like a painter or poet, is a maker of patterns and these must be beautiful. Beauty is the first test: there is no permanent place in the world for ugly mathematics." But, says Mordell, "I do not think that Hardy's work is characterized by beauty. It is distinguished more by his insight, his generality, and the power he displays in carrying out his ideas. Many of the results that he obtains are very important indeed, but the proofs are often long and require concentrated attention, and this may blunt one's feelings even if the ideas are beautiful."[33]

Mordell ends by refuting Hardy's dictum in which he said, "Mathematics is not a contemplative subject." Replies Mordell: "What does he mean when he says mathematics is not a contemplative subject? Many people can derive a great deal of pleasure from the contemplation of mathematics, e.g., from the beauty of its proofs, the importance of its results, and the history of its development. But alas, apparently not Hardy."[34]

Opinions about aging in mathematics vary broadly. Following are a few examples. Members of the Bourbaki collective withdrew from active group participation at 50. Albert Einstein said, "A person who has not made his great contribution to science before the age of thirty will never do so."[35] The French number theorist André Weil wrote, "There are examples to show that in mathematics an old person can do useful work, even inspired work, but they are rare and each case fills us with wonder and admiration."[36] At Felix Klein's 50th birthday party, he whispered to Grace Chisholm Young: "Ah, I envy you. You are in the happy age of productivity. When everyone begins to speak well of you, you are on the downward road."[37]

In an article in *The New Yorker*, "Mathematics and Creativity," Alfred Adler wrote: "consuming commitment can rarely be continued into middle and old age, and mathematicians, after a time, do minor work. In addition, mathematics is continually generating new concepts which seem profound to the older men and must be painstakingly studied and learned. The young mathematicians absorb these concepts in their university studies and find them simple. What is agonizingly difficult for their teachers appears only natural to them. The students begin where the teachers have stopped, the teachers become scholarly observers."[38]

L. E. J. Brouwer (1881–1966), one of the founders of algebraic topology and the great leader of the intuitionist school on the foundations of mathematics, was 24 when he wrote about older academics, "There are others who do not know when to stop, who keep on and on until they go mad. They grow bald, short-sighted and fat, their stomachs stop working, and moaning with asthma and gastric trouble they fancy that in this way equilibrium is within reach and almost reached. So much for science, the last flower and ossification of culture."[39]

On the other hand, Joseph Dauben, the biographer of the Israeli-American logician Abraham Robinson, wrote, "[Robinson] was always pleased to dispel the myth that the best mathematicians were under thirty and that a mathematician did her or his best work early, at the very start of one's career. As a striking counterexample, Robinson's best mathematics was only beginning to reap the benefits of his wide experience when, suddenly at the age of fifty-five, he died."[40]

Abraham de Moivre (1667–1754) found his presumably most important result when he was 66: "the local central limit theorem." It is reported in all seriousness that de Moivre correctly predicted the day of his own death. Noting that he was sleeping 15 minutes longer each day, de Moivre surmised that he would die on the day he would sleep for 24 hours. A simple mathematical calculation quickly yielded the date, November 27, 1754. He did indeed pass away on that day.[41]

The English algebraist J. J. Sylvester pointed out that Leibniz, Newton, Euler, Lagrange, Laplace, Gauss, Plato, Archimedes, and Pythagoras all were productive until their seventies or eighties. "The mathematician lives long and lives young," he wrote. "The wings of the soul do not early drop off, nor do its pores become clogged with the earthy particles blown from the dusty highways of vulgar life." Sylvester himself was in his 82nd year, in 1896, when he "found a new enthusiasm and blazed up again over the theory of compound partitions and Goldbach's conjecture."[42]

A Survey on Mathematical Aging[43]

Who is right, Hardy or Mordell? The question intrigued Reuben Hersh who decided to check up on whether mathematics really is a young man's game. From the American Mathematical Society membership directory, he chose 250 names of men and women he had known somewhere, at some time.

They were mostly Americans but also included were a few Canadian, Swedish, French, Israeli, and Japanese mathematicians who had spent time in the United States. Reflecting his training and experience, the mailing list was heavy on workers in differential equations: theoretical (both ordinary and partial), applied, numerical, or stochastic. There were also stochastic processors, and a scattering of logicians, algebraists, topologists, geometers, and statisticians. Mathematical specialty should not make much difference for the questions. Of course, the anonymity of these responses was respected.

There were several clusters of questions. Some dealt with the value of the respondents' early work in contrast with later achievements. Did they experience less enthusiasm or a sense of loss as a consequence of aging? Did the respondent shift emphasis from research to teaching or administration? How did aging affect the status of the respondents in their departments and in the larger mathematical community? Did the mathematicians who answered have suggestions for individuals or institutions when confronted with the challenges of aging?

There were 66 replies, which is considered a good response rate. They came from 23 states, 3 Canadian provinces, Sweden, and Israel. Ages ranged from 60 to 92. There is no claim that this choice of 250 was typical, let alone random. And the 66 of 250 who answered are certainly not typical. They are biased toward people who answer questionnaires, who like to hear from an old acquaintance, who are willing to consider some possibly painful issues, and who aren't too unhappy or ashamed of their lot in mathematical life. The people who don't respond to questionnaires are like the dark matter of the cosmos; we know they are out there, but we can only guess what they look like.

Two respondents knew of earlier unpublished surveys by famous mathematicians. George Mackey of Harvard did a study of 50 leading mathematicians and concluded that on average their best work was done in their late thirties. The prominent topologist Gail Young, a student of Robert Lee Moore, did a study of people who matured very young in mathematics. He found that they generally burned out early. Young felt there was a fairly constant period during which a person could do very creative work. Some had their period earlier; others had it later.

Hersh's questionnaire invited recipients to reveal as much as they liked about their current and past situations. Their answers yield a glimpse of how this sample of mathematicians view their lives in mathematics. Such responses don't submit readily to tabulation, much less to statistical analysis. These reflections are of interest in their diversity and how they depict individual points of view.

Most of the respondents were satisfied with their life situations. Relevant is a report by S. S. Taylor on retirees (not restricted as to field) at the universities of New Mexico and Rhode Island in the United States and Bath and Sussex in England. Reassuringly, perhaps surprisingly, 98 percent of the New Mexico retirees, 97 percent of the Rhode Island retirees, and 84 percent of the English retirees told Taylor they were "reasonably satisfied" or even "very satisfied" with retirement. Two-thirds of the American respondents told Taylor they received the same or a higher income as before retirement.[44] Most of Hersh's respondents say they continue research after retirement. Some think their recent research is their best ever. Some say they're doing what they're interested in, uninhibited by the judgment of the math community.

Survey Results from Hersh's Questionnaire

The responses are organized into six groups. Excerpts from some answers appear in more than one group.

1. Mathematicians report losing their edge once past youth.
2. Mathematicians may be as good or better in later years.
3. Symptoms of age and coping strategies.

4. Penalties for aging and for following one's own bent.

5. Advice for aging mathematicians.

6. Advice for mathematics departments.

Many responses are hard to classify. For example, one mathematician is reported to have just published a very good paper at age 75, though he complains bitterly about not being able to do good research any longer. Does this belong in the first group, the second group, or both? Some painful experiences in the fourth group contradict some advice in the fifth group. Many respondents say, "Follow your own bent, regardless of outside pressure"; but many respondents report penalties for doing that. Some respondents don't give their age; for a few, Hersh was unable to identify their geographical location. The following excerpts are taken from longer messages.

1. Mathematicians Report Losing their Edge Once Past Youth

There are some reports of sad or difficult experiences. A 72-year-old analyst in California confessed, "My zest is fine, but capacity much diminished before age 56. Age and alcohol and depression." A California geometer explained, "The field of mathematics moves very fast. The pace has been quite extraordinary in the past 50 years. Just trying to keep up in one's specialty requires many hours of effort. One doesn't feel comfortable doing the same old thing. Some great mathematicians have been unable to handle this. When a decent problem comes along which seems accessible, I'm eager to jump in. The trouble is that the frontier is moving so fast. It's not that we give up research mathematics, research mathematics gets away from us."

And a 79-year-old analyst in Maryland wrote, "At around 56 I had lost whatever originality I once possessed. But not the desire to learn and try." A 62-year-old friend in Louisiana, also an analyst, wrote, "I used to work late at night, but now I'm too exhausted to do more than make calendar entries and clean up my study." And a 71-year-old applied mathematician in Rhode Island said, "Clearly at my age I can't keep up with the best younger people. Some old-timers have looked foolish in their later research efforts. My hope is at least to avoid that."

The well-known number theorist Ivan Niven was quoted, "As you get older you know too much. You have all these methods, and you try all the combinations and variations you can think of. You're running down the old tracks and nothing works."[45]

And, in explanation of these stories of loss, one respondent, whose age and location are unknown, wrote, "Mathematics tends to be introverted, with unbalanced expenditure of mental energy. As one grows older there is desire for other forms of expression, which dilutes the intensity to solve problems. 'What does it all mean?' is asked more often, which also can slow down progress."

And a famous Swedish analyst, at age 80, explained, "Aging has two sides—your own age, and the age and aging of your subject and your contributions. This aging is brought about by the work of younger competitors."

On the other hand, some presented a different view.

2. Mathematicians May Be as Good or Better in Their Later Years

This turns out to be the opinion of most women mathematicians. Claudia Henrion interviewed half a dozen leading female U.S. mathematicians,[46] and they told her that they had done their best work in their thirties, forties, even fifties. Lenore Blum returned to mathematical research in her forties, after years of teaching at Mills College and involvement in national programs to promote women in mathematics.

Mary Ellen Rudin, who managed to stay professionally active even while raising four children, reports that she is doing some of her best work in her fifties and sixties, now that most of the children are grown. She worked part-time as a lecturer until she was almost 50, when the University of Wisconsin promoted her from a lecturer to a full professor. Rudin said, "I don't think most people's best work will be done by the time they're thirty, and certainly my best work wasn't done until I was fifty-five years old."[47] Others may follow Mary Ellen Rudin's path and work part time for a period to balance having children with mathematical research.

The prominent logician Marian Pour-El told Henrion in an interview, "I've never felt that you're over the hill if you're in your late thirties. I think I've done my best work later on, by a long shot."[48]

Joan Birman, a topologist at Columbia University–Barnard College, did not get her Ph.D. until she was 40 years old. Birman focused better on math after the issues of marriage were settled, her children were older, etc. "I think doing mathematics when you're enthusiastic is important—not your age."[49] In fact, most of the women Henrion interviewed found that their work improved as they got older.

A 62-year-old female probabilist in Hersh's sample wrote, "Men age faster than we girls. It makes up for them being bullies earlier. How to pep them up? I try. . . . People whose lives are fairly stable and satisfactory keep going a lot better. One of my colleagues gave up research at 42 when his marriage broke up. Another similar at 48."

A male member from the sample emphasized collaboration as a source of long-term productivity. This 71-year-old analyst–applied mathematician in Rhode Island wrote, "Since I became emeritus in July 1995 my research has increased. Most of it is joint with former students and postdocs. The mathematical tools are ones I've used before. This is probably typical. It is a great relief to shed 9 years of department chairmanship, too many committees, and obligations to seek external grants. I no longer attend department meetings."

Then there was the Illinois analyst, aged 69, who told Hersh: "Some of my best work was done after age 47. Possible motivations were a bad spell of drinking and divorce from 1974 to 1977, and prostate cancer treated successfully by radiation and implants. After such trauma I tend to over-accomplish."

A Colorado applied mathematician thought, "Young guys may luck out but often only when someone older points the way," and another respondent put it this way: "The young may find gold but cannot read the land; the older have familiarity with the landscape, which guides them to where to dig." Best of all was the comment from a 70-year-old Ohio numerical analyst: "Recently a friend compared me with Brahms, who turned out great works throughout his life! I hope to live up to the praise."

3. Symptoms of Aging, and Coping Strategies

In the responses to the survey there were many suggestions to address the challenges of aging. An applied mathematician of age 70 wrote, "As I age my memory declines, making it more difficult to keep in mind all the threads of a complicated situation. Also my computational abilities decline. I take longer to get through a routine calculation, and make more mistakes. I catch mistakes by my sense of what seems right, rather than by repeating computations. On the other hand, I'm more canny in developing effective research strategy, and more daring in carrying it out. . . . I have an intellectual home with a small but active worldwide community of scholars with similar interests."

And a 74-year-old numerical analyst in California said, "I toy with retiring at the end of this year. I am nervous about it, but clearly recognize the diminishment of my ability to do first-rate research. The main cause is inability to stick with messy detailed manipulations. In the past I could calculate for hours, but now I shy away from such grunge. I still have plenty of things to work on, but I pick them more carefully."

Another numerical analyst, who shares his time between California and Sweden, bluntly declared:

Getting old is a pain. I still do decent mathematics. However, what I do is very much related to my previous work. I do not jump into a new field, because I have not the same intuition as earlier to 'know' it will lead to something. Everything takes much longer to complete and I make more mistakes, or better,

I do not know immediately when the result is wrong. So I have to check much more carefully. I have been a good thesis advisor, which I enjoyed very much. Former students still speak to me, and I still work with them. But I have no students anymore, because I cannot be sure I will be around in 4 years. Also, young people should work with young people on "modern" problems. There can be one advantage with old age. If one is lucky and in balance with oneself, one can look at the world as an independent observer."

A male topologist, 76 years old, of British origin and now living in New York State, analyzed the situation carefully:

The principal obstacles to continuing research are: (a) Research requires energy, and this is in increasingly short supply. (b) Research requires keeping up with the literature, and this becomes difficult as one's mental and physical energy declines. (c) Good research requires breadth and flexibility, but the tendency as one ages is to concentrate on a narrow path, dominated by what one has always done, and knows well. Collaboration is essential in maintaining research activity. I have tended to collaborate with juniors, since very many of my collaborators have been my students. The younger partner provides energy and awareness of what is currently a "hot topic"; the older provides perhaps greater familiarity with the history of the topic and a larger battery of available methods.

(Gian-Carlo Rota made a similar remark, "At my age, the collaborator is a necessity."[50])

And a 57-year-old logician in Indiana also had an explanation. "There are many useful things someone with mathematical ability can do. But education and research rewards do not encourage people to branch out and explore. They get stuck in the frontiers of their narrow specialty. The going gets rough when they no longer have the ability or willingness for the concentrated effort to do really complicated technical work. I am still able to do this if I get away for a couple of weeks, but at home commitments to family and work preclude that concentration. It does get harder as you get older, from aging but also from accumulation of other responsibilities and interests."

The shortest and sweetest lines came from an applied mathematician in Rhode Island, aged 71, who described his coping strategy. "My wife and I have been happily married for forty-four years; that's extremely important. Our garden takes a major part of our time in the growing season."

4. Penalties for Independence

Several respondents reported penalties for following one's own path. These comments suggest the need for some soul searching by mathematics departments! Thus, an applied mathematician in British Columbia reported, "My recent work is more interesting and valuable. Math community isn't interested. Ecology community is." And a 68-year-old number theorist in Minnesota said, "My best paper was never referred to in the later literature. I tell myself this shows it said the last word on its subject." And a 57-year-old logician in Indiana admitted, "I did feel a loss when what I was doing was not valued by the mathematical community. It took a while for me to value it for its own sake."

And a 72-year-old analyst in California commented, "The mathematical community lost interest in my work when fashions changed and I didn't. After a period as chairman when I was 40, I lost influence in the department." And still one more replied, "By following my bliss, I gave up my opportunity to get to full professor. My most valuable professional achievements are not appreciated by the leadership in my department." And finally, in this vein, is the response "Some of my best research has been in recent years, yet I have been getting smaller pay raises and have less influence in my department The situation of some of my contemporaries is even more egregious. Mathematics departments and organizations don't pay attention to the older members of the profession. My department treats our retirees shabbily: we give them a 'gold watch' when they retire, then forget about them."

An applied mathematician in Alberta reported that he helps out with administrative chores. "The value of my research = quite high, the interest by others = quite low. The math community doesn't pay attention to most mathematicians' work. . . . I am called on a lot to do diplomatic or administrative jobs. . . . I am not a very able administrator, but compared to the great majority of mathematicians, I am an administrative genius."

But here are a couple of comments on the plus side of the story. An 80-year-old very well-known measure theorist in California said, "I have been treated well. I still have my office, 10 years after retirement." And similarly, a 66-year-old Swedish numerical analyst reported, "My department has treated me well. I still have an office and they pay me a small amount for looking after some graduate students. My research is worth more to me than to the department, so there is no strong reason why they should actively support it."

5. Advice to Aging Mathematicians

Many respondents had strong advice for their fellow retirees. The great analyst-topologist Isadore (Izzy) Singer was quoted by one of the respondents: "Keep the pencil moving." This theme resonated in other responses, such as that from the 68-year-old Swedish analyst who said, "Keep working, do not hide behind administrative duties." And a 67-year-old logician wrote, "Work hard, and have several problems to work on." And a 65-year-old geometer in New York advised: "Don't stop. Once you do, it's hard to get back. It's not just the field that changes, but you change." "First and foremost, you need a deep love of the subject" (Analyst, Alberta, age 60.)

There was also more practical and specific advice. "It is important to have an office," said a Swedish numerical analyst, age 66. And an applied mathematician in Utah said, "Stay away from administration. It wears away your creativity, and is a real plague." A 59-year-old probabilist in Illinois claimed that

"Hardy advised people to do research in a prone position, so that more blood flows to the head."

A 71-year-old harmonic analyst in Sweden, who had been one of Hersh's professors when he took advanced calculus at the Courant Institute in 1957, wrote, "Maintain contact with younger colleagues and students. . . . Whenever anyone asks you a mathematical question, devote at least 15 minutes to it, even if it's 'not [your] field.'. . . Try to maintain high ethical standards in this competitive profession. I like to think of mathematics as a collective enterprise. We contribute even by being attentive spectators and consumers of the constant outpouring of new ideas. (In the opposite view, a career in mathematics is an ego trip like downhill skiing, reserved for the youngest and strongest, where only those who break records matter.)"

Finally, we are reminded of the basics. "Always remember, research should be fun. If it becomes too competitive and loses its pleasure, give it up. Don't take your research or yourself too seriously! I have been blessed with a good sense of humor, but how could one suggest this to others?" (Numerical analyst, British Columbia, age 54.)

"I do what I can do, and enjoy every minute of it. The mathematical community has as little awareness of me as I of them. Constantly learn new things! Do mathematics just for the fun of it!" (Number theorist, New York, age 77.) "Stop growing older. Keep having fun. And have a beer, on me!" (Analyst, California, age 83.)

Now we go to the last category of responses to our questionnaire.

6. Advice for the Mathematics Community

A 74-year old applied mathematician from California proposed a radical shake-up in the usual career path for mathematicians. "I have believed for a long time that a lifetime appointment to research in mathematics, with incidental teaching, is a mistake," he wrote. "A person's abilities, skills, and interests change. I've often talked about a career path involving research at an early age, say 25–35, writing and teaching at a non-research university thereafter, perhaps with involvement in pre-university mathematics such as teacher training or high school teaching. . . . Shorten the time to get a Ph.D. so that people can start research earlier, as in England. Shorten the undergraduate school time for talented people."

A 77-year-old number theorist in New York had a very sharp criticism of universities. "For me, a big issue is why retirement is synonymous with severance from all academic activities. The University is the one organism that consciously believes there is nothing to learn from the past."

"Encourage individuals to take chances and follow their true interests," said an applied mathematician in Montana (age 70). "I don't think anyone should tell us when to give up," said a Californian, age 72. And "a good library, some stimulating colleagues and freedom from too many onerous chores" were all that a 60-year-old Alberta, Canada, analyst required.

Shabby treatment of aging professors is not special to mathematics. After retiring from the Columbia economics department, William Vickrey got a Nobel Prize for work on transportation economics. A *New York Times* reporter found him in a tiny office far from his department. Vickrey was grateful that after retirement Columbia had allowed him any office at all. Perhaps after being written up in the *Times* he would have been granted a better office, but, sadly and unexpectedly, he died a few days later.

A fuller account is Bollobás' description of Littlewood's retirement at Cambridge:

> In 1950 at the statutory age of 65, he retired and became an Emeritus Professor. The Faculty Board realized that it would be madness to lose the services of the most eminent mathematician in England, so they wrote to the General Board: "Professor Littlewood is not only exceptionally eminent but is still at the height

of his powers. The loss of his teaching would be irreparable and it is avoidable. Permission is requested to pay a fee of the order of 100 pounds for each term course of lectures." The response: 15 pounds per term, the fee paid to an apprentice giving his first course as a try out to a class of 2 or 3. So he gave courses at 15 pounds for 4 years. He tried to stop once but there was a cry of distress. At the same time he was declining lucrative offers from the United States."[51]

Conclusions

The responses varied. There are tremendous differences in how mathematicians age. Until we find a consensus about which advances are "major," we can't refute Hardy's claim that no major advance has been made by a mathematician over 50. But his slogan, "Mathematics is a young man's game," is misleading, even harmful. So far as it may discourage people from mathematics when they're no longer young, it's unjustified and destructive.

Some of these answers are advice to aging mathematicians. Most important, the respondents say, "Don't stop. Keep working." Some of them found greater pleasure in their work after retirement, when they felt free to work on problems not considered important or prestigious or urgent by other people. Many of them were isolated from former colleagues, but some succeeded in developing new collaborations, especially with younger people. As expectations about age change, many mathematicians find new problems and interests past the arbitrary dividing line of retirement.

Some of the comments we quoted are important for department policy. Many of them make clear the importance of strong mathematical communities. When older or retired mathematicians are cut off from their departments, this amounts to making insiders into outsiders. It weakens the departmental community.

Older and retired mathematicians are an underutilized resource for the mathematics community. Departments can always use extra hands. Undergraduate advising is often understaffed. Has anybody asked, Are there emeriti who enjoy advising? If there's no librarian on duty in the math library, is there an emeritus who would serve? There's always too much committee work. Is there an emeritus with years of service on the undergraduate committee or the master's exam committee? Might he or she have something to contribute there?

Although aging brings losses in memory and computing ability to most (but not all) mathematicians, these may be compensated for by a broader perspective and mature judgment. Older and retired mathematicians are certainly a valuable part of the mathematical community, and they should be recognized as such.

Individuals differ greatly in the ways in which they respond to the challenges of aging and retirement. But interestingly, there are differences between men and women in the trajectory of their productive lives. While many men see their thirties and forties as a particularly productive period and one in which they focus keenly on their work to the exclusion of other interests and responsibilities, women develop more slowly and hit their stride past their childbearing years. In both groups, aging is most successful for those who have established a balance among work, personal life, and broader interests. One of the difficulties of aging is the isolation from colleagues and former students. Lack of institutional support and outmoded departmental practices frequently contribute to such isolation. Men and women who are able to reach out to others, who continue to be valued by their community, and who partner with younger colleagues express the greatest satisfaction with their lives past retirement. This continued engagement of older mathematicians enriches the field and provides a sense of mathematical life in all its fullness to the younger members of the profession.

Bibliography

Adler, A. (1972). Mathematics and creativity. *The New Yorker*, February 19, 1972. Reprinted in Timothy Ferris (Ed.) (1993). *The world treasury of physics, astronomy and mathematics*. Boston, MA: Back Bag Books, p. 435.

Albers, D., & Alexanderson. G. (1991). A conversation with Ivan Niven. *College Mathematics Journal 22*(5), 371–402.

Bell, E. T. (1937). *Men of mathematics*. New York: Simon and Schuster.

Case, B. A., & Leggett, A. M. (Eds.) (2005). *Complexities: Women in mathematics*. Princeton, N.J.: Princeton University Press.

Dauben, J. (1995). *Abraham Robinson: The creation of nonstandard analysis: A personal and mathematical odyssey*. Princeton. N.J.: Princeton University Press.

Einstein, A. (1942). Quoted by Stern. ref. S. Brodetsky, *Nature 150* (1942), 699, as quoted in C. W. Adams (1946), The age at which scientists do their best work, *Isis 361*166–169.

Germain-Gauss correspondence. <http://wwwgroups.dcs.stand .ac.uk/%7Ehistory/Mathematicians/Gauss.html>

Gray, M. W. (1976). Sophie Germain: A bicentennial appreciation, *AWM Newsletter* 6(6), 10–14. [Reprinted in *Complexities: Women in Mathematics*, Bettye Anne Case & Anne Leggett (Eds.). Princeton, N.J.: Princeton University Press, 2005, 68–74.]

Gray, M. W. (1987). Sophie Germain. In L. S. Grinstein & P. J. Campbell (Eds.). *Women of mathematics: A biobibliographic sourcebook*. Westport, Conn.: Greenwood Press, pp. 47–56. This article contains an excellent bibliography of works written about Sophie Germain.

Hardy, G. H. (1940). *A mathematician's apology*. New York: Cambridge University Press.

Henrion, C. (1991). Merging and emerging lives: Women in mathematics. *Notices* of the *American Mathematical Society 38*(7), 724–729.

Henrion, C. (1997). *Women in mathematics: The addition of differ-ence*. Bloomington, Ind.: Indiana University Press.

Hersh, R. (2001). Mathematical menopause, or, a young man's game? *Mathematical Intelligencer* 23(3), 52–60.

Hyde, J. (2005). The gender similarities hypothesis. *American Psychologist 60*, 581–592.

James, I. (2002). *Remarkable mathematicians*. Cambridge: Cambridge University Press.

James , I. (2009). *Driven to innovate: A century of Jewish mathematicians and physicists*. Oxford: Peter Lang.

Kovalevskaya, S., Kochina, P. Y., & Stillman, B. (1978). *A Russian childhood*. New York: Springer, p. 35.

Kummer <http://www-groups.dcs.st-and.ac.uk/%7Ehistory/Mathematicians/Kummer.html>

LaGrange <http://www-groups.dcs.st-and.ac.uk/%7Ehistory/Mathematicians/Lagrange.html>

Legendre-Germain correspondence <http://www-groups.dcs.st-and.ac.uk/%7Ehistory/Mathematicians/Legendre.html>

Lehman, H. C. (1953). *Age and achievement*. Princeton, N.J.: Princeton University Press.

Littlewood, J. (1986). *Littlewood's miscellany*. Preface by Béla Bollobás. New York: Cambridge University Press.

Mordell, L. J. (1970). Hardy's *A mathematician's apology*. *American Mathematical Monthly 77*, 831–836.

Murray, Margaret (2000). *Women becoming mathematicians*. Cambridge, Mass.: Massachusetts Institute of Technology.

Noether, G. E. (1987). Emmy Noether (1882–1935). In Louise S. Grinstein & Paul J. Campbell. *Women of mathematics: A biobibliographic sourcebook*. New York: Greenwood Press, pp. 165–170.

Perl, T., & Morrow, C. (Eds.) (1998). *Notable women in mathematics*. Westport, Conn.: Greenwood Press.

Reid, C. (1970). *Hilbert*. New York: Springer-Verlag.

Reid. C. (1986). *Hilbert-Courant*. New York: Springer-Verlag.

Senechal, M. (2007). Hardy as mentor. *Mathematical Intelligencer* 29(1), 16–23.

Smolin, L. (2006). *The trouble with physics*. London: Houghton-Mifflin: Penguin.

Stern, N. (1978). Age and achievement in mathematics: A case-study in the sociology of science, *Social studies of science*, vol. 8. Beverly Hills, Calif.: Sage, pp. 127–140.

Taylor, S. S. (1999). *Research dialogues of the TIM-CREF*, no. 62.

van Stigt, W. P. (1990). *Brouwer's intuitionism*. Amsterdam: North-Holland.

Weierstrass <http://www-groups.dcs.stand.ac.uk/%7Ehistory/Mathematicians/Weierstrass.html>

Weil, A. (1950). The future of mathematics. *American Mathematical Monthly 57*, 296.

Weyl, H. (1935). Emmy Noether. *Scripta Mathematica 3*(3), 201–220.

Wiegand, S. (1977). Grace Chisholm Young. *Association for Women in Mathematics Newsletter 7*, 6.

+8+

The Teaching of Mathematics: Fierce or Friendly?

How is mathematical practice shaped by the social context in which it takes place? How are mathematicians affected by the beliefs, biases, and values of the subculture in which they work and teach?

In the previous chapter we wrote about the impact of age and gender on the ways in which mathematicians develop and sustain their lives. We emphasized the importance of balance and social connections as sources of support when researchers face societal stereotypes. Some may even find in mathematics an escape from social challenges, from "everyday life, with its painful harshness and wretched dreariness, and from the fetters of one's shifting desire."[1] But more often, one's mathematical life is intimately entwined with the life and conditions of the larger society. The impact of a mathematician's early socialization and its attending values and world view can be so pervasive that it influences the researcher's lifelong behavior.

In this chapter we tell about two U.S. mathematicians whose lives and work were deeply embedded in their time and place. For most of America's history, the lives of Black people and White people were sharply segregated, and yet, paradoxically they were intimately entwined. We have chosen two mathematicians whose character and practice embodied this paradox. Their backgrounds were as different as one could imagine.

Robert Lee Moore and Clarence Francis Stephens

Robert Lee Moore (1882–1974) was Texas-born. His father, Charles, although Connecticut-born, had made himself a southerner and served in the Civil War as a Confederate volunteer, surviving the battles of Shiloh, Vicksburg, and Chattanooga. On his mother's side Moore was related to Jefferson Davis, the President of the Confederacy. In his later years "by a masterwork of genealogical detective work, he produced a chart of lineage that would embrace two American presidents, the president of the Confederacy and three European royal houses."[2]

Clarence Francis Stephens was born in Macon, Georgia, in 1917, the son of Sam Stephens, a chef and railroad worker, and Jeannette Morehead Stephens. He was the fifth of six children. His mother died when he was 2 and his father when he was 8, so he and his brothers and sisters went to live with their grandmother. She died two years later, and then the children were separated and sent to live with different relatives. Stephens lived with his great aunt Sarah in Harrisburg, North Carolina. There were no high schools in Harrisburg, so when he was 13, his oldest sister Irene arranged for him to go to a boarding school in Irmo, South Carolina. She paid the first year's tuition, but Stephens himself earned his subsequent tuition, working on the school farm and in the kitchen and cleaning classrooms. After a placement test he was put into the 8th grade alongside students who were over 20. He played football and baseball, held the lead role in his senior play, and was elected class president.[3]

He earned a B.S. in mathematics from a traditionally Black institution (TBI), Johnson C. Smith University in Charlotte, North Carolina, while supporting himself as a deliveryman for a local drugstore at 6 dollars a week. Then he went to the University of Michigan for a master's degree, expecting to become a high school teacher. But Professor George Rainich encouraged him to continue for a Ph.D. At Michigan, at that time, "there were no teaching assistant positions for African Americans," so he supported himself by waiting tables.[4]

In September 1940, before finishing his Ph.D. at Ann Arbor, Stephens began teaching at another traditionally Black institution, Prairie View A. & M. in Texas. After Pearl Harbor he joined the U.S. Navy as a teacher specialist. He finished his Ph.D. in the spring of 1943, returned to teaching at Prairie View A. & M., and stayed through August 1947. Then he applied for a position at Morgan State College in Baltimore, in order to be near the math research library at Johns Hopkins. "To his surprise and initial dismay, he was offered the position of Chair of the department, which would put him in charge of a department of mathematicians, all older than he, at the tender academic age of 30. His desire to live in Baltimore overcame his concerns, and he accepted [in September 1947.]"[5]

Like Johnson C. Smith and Prairie View, Morgan State was also a traditionally Black institution. James A. Donaldson, mathematics professor and dean at Howard University, has written: "After termination of the U.S. Civil War in 1865, free Black people and newly freed Black people, fortified by hope and quiet determination, struggled to prepare themselves in every way for full membership in society. Consequently this period, shortly before and after 1865, saw the founding of several educational institutions (Lincoln University in Pennsylvania, Wilberforce, Howard, Shaw, Johnson C. Smith and others which will be called traditionally Black institutions or TBI's) with the goal of providing higher educational opportunities for newly freed Black people and other people of African descent."[6]

Moore and Stephens eventually developed two unique, radically different styles of college-level mathematics education. The decentralized nature of the American educational system permits both Moore's and Stephens' traditions to survive. Moore's method is more widely known than Stephens' (which is generally known as the Potsdam model). Yet, in somewhat different circles than the Moore method, the Stephens method has also received a lot of recognition. The contrast between these two traditions reveals a great deal about the United States, its math pedagogy, its ideologies, and its racism. We will first describe

Moore and his method and then return to the career of Clarence Stephens, first at Prairie View and Morgan State and then at the racially integrated college at Potsdam, New York.

Moore and His Method

In 1898 Moore became a student and protégé of George Bruce Halsted at the University of Texas. Halsted "could point with pardonable pride to the fact that the rolls of the College of New Jersey, at Princeton, bore not only the names of his brother and himself, but also those of his father, an uncle, his grand-father, a great-uncle and his great-grandfather."[7] Halsted was the first doctoral student of the great algebraist J. J. Sylvester at Johns Hopkins, in the first math doctoral program in the United States. Later, Halsted pioneered teaching and writing about non-Euclidean geometry. He even went to Hungary and found the grave of János Bolyai. The Hungarian authorities were surprised that this obscure grave site interested a professor from Texas.

As a professor in Texas, Halsted engaged in legendary fights with the university's board of regents. His star pupil was young Robert Lee Moore. He mentored Moore through a bachelor's degree and then sent him on to the University of Chicago. At Chicago Moore became friends with fellow student Oswald Veblen, who would become a powerful professor, first at Princeton University and then at the Institute for Advanced Study. As students, Veblen and Moore became interested in the axiomatic foundations of geometry, a popular research subject after David Hilbert published his famous *Foundations of Geometry*. After his studies at Chicago, Moore taught at Northwestern, at the University of Pennsylvania, and finally back at the University of Texas in Austin. "With his snowy white hair immaculately combed, his piercing blue eyes always seeking exciting new proofs to complex problems, and his well-muscled boxer's physique clad in dark three-piece suits and old-fashioned, hand-made laced-up black boots, he was a commanding presence on the campus of The University of Texas for 49 years."[8] (The "well-muscled boxer's physique" is no mere metaphor. In his young days Moore did enjoy and excel at the sport of amateur boxing.)

At Pennsylvania Moore supervised the thesis of John R. Kline (1891–1955), who remained there and ultimately was department chairman from 1933 to 1954. Kline also was secretary of the American Mathematical Society (AMS) from 1936 to 1950—we will say more about him in a little while.

Moore early turned from geometry to what he called "point-set theory." This was part of point-set topology, which was a very active field in the early 20th century. A central question was the question of "metrization." A topological space is one where there are "neighborhoods." A topological space is called a "metric space" if the neighborhoods are defined in terms of a "metric" or "distance." For instance, an ordinary surface such as a sphere becomes a metric space if distance between points is measured along the "geodesic," the shortest curve on the surface connecting them. But some topological spaces do not permit a metric or distance to be defined between points. What extra conditions on a topological space will permit a metric to be defined?[9]

Moore's fame rests less on his own research than on his teaching method, and on the amazing number and accomplishments of his students. Through his 50 Ph.D. students he has 1678 doctoral descendents! Three of them followed him as president of the American Mathematical Society; three others became vice-president; and another served as secretary of the AMS. Moreover, five served as president of the Mathematical Association of America (MAA), and three, like Moore, became members of the National Academy of Sciences.

The essence of the Moore method is easy to describe. He recruited students beginning their first year so that he could control their mathematical education from the very start. They were not permitted to have any previous knowledge of topology. Also, they were forbidden to read any books or articles about it. Also, they were even forbidden to talk about it outside of class.

Figure 8-1. Robert Lee Moore. Courtesy of
Dolph Briscoe Center for American History, The
University of Texas at Austin. Identifier: di_05554.
Title: R. L. Moore and Mike Profit. Date: 1970/01.
Source: Halmos (Paul) Photograph Collection

At the first meeting of his graduate class in point-set theory, Moore would present his axioms for point sets and give the statement of theorem 1, without proof or hints. (Perhaps he might also state theorems 2 and 3.) That's it! Nothing else would happen, in that class or the next, until one of the students claimed he or she could prove theorem 1. The student then presented the proof at the blackboard. The class and the professor reacted to the proof. Eventually, theorem 1 was proved, and the class was ready for some student to propose a proof of theorem 2.

This method is reminiscent of a well-known old method of teaching swimming called "sink or swim." Moore persisted in this teaching method at Austin until 1969, when, at age 87, he was finally, with great difficulty, forced to retire.

What happened in class when nobody was ready to offer a proof? Moore did not take the opportunity to lecture on some mathematical topic. Not at all. Instead, he simply chatted about any casual matter that happened to come up. In his chit-chat he sometimes expressed low opinions about women, about Black people, and about Jews. When his student Gayle Ball told him that her husband, Joe, was planning to go to a northern university after graduation, Moore warned Ms. Ball that she might "have to, perhaps, get on a bus and sit down next to a person of color. He thought this would be very offensive. . . . He said, 'What would you do, Mrs. Ball, if this happened to you?' She answered, "I would say: 'How do you do?'" Moore's reaction is not recorded.[10]

Another of his students, Mary Ellen Estill, later known to the world as Mary Ellen Rudin, would direct 16 doctoral theses and become a vice-president of the American Mathematical Society. In her interview for the book *More Mathematical People*, she talked frankly about her mentor Robert Lee Moore.

MP. So you had a course from Moore every semester?

Rudin. Every single semester during my entire career at the University of Texas. I'm a mathematician because Moore caught me and demanded that I become a mathematician. He schooled me and pushed me at just the right rate. He always looked for people who had not been influenced by other mathematical experiences, and he caught me before I had been subjected to influence of any kind. I was pure, unadulterated. He almost never got anybody like that. I'm a child of Moore. I was always conscious of being maneuvered by him. I hated being maneuvered . . . he maneuvered you in order to build your ego. He built your confidence that you could do anything. I have that total confidence to this day . . . having failed 5,000 times doesn't seem to make me any less confident. . . . We were a fantastic class. [The class of '45 included, besides Mary Ellen Rudin, R H Bing, R. D. Anderson,

Figure 8-2. Mary Ellen (Estill) Rudin. Source: *More Mathematical People: Contemporary Conversations.* Eds. Donald J. Albers, Gerald L. Alexanderson, and Constance Reid.

Gail Young, and Ed Moise!] Each of us could eat the others up. Moore did this to you. He somehow built up your ego and your competitiveness. Actually in our group there was another, a sixth, whom we killed off right away. He was a very smart guy—I think he went into computer science eventually—but he wasn't strong enough to compete with the rest of us. Moore always began with him and then let one of us show him how to solve the problem correctly. And boy, did this work badly for him! It builds your ego to be able to do a problem when someone else can't but it destroys that person's ego. I never liked that feature of Moore's classes. Yet I participated in it. Very often in the undergraduate class, I mean, I was the killer. He used me that way, and I was conscious of being used that way. . . . He viewed his two earlier women students as failures, and he didn't hesitate to tell me about them in great detail, so I would realize he didn't want to have another failure as a woman. . . . I only knew the mathematics that Moore fed me. At the time I wrote my thesis I had never in my life seen a single mathematical paper! I was pure and unadulterated. The mathematical language that he used was his own. I didn't know how mathematical words were used at all.

MP. How did you feel about your mathematical education later?

Rudin. I really resented it, I admit. I felt cheated because, although I had a Ph.D., I had never really been to graduate school. I hadn't learned any of the things that people ordinarily learn when they go to graduate school. . . I didn't even know what an analytic function was.

MP. What feelings toward Moore, as a person, did you develop over time?

Rudin. Oh, I had a very warm, enthusiastic feeling for him, although I also had very negative feelings. I was conscious of both levels. I was aware that he was bigoted—he was—but I also was aware that he played the role of a bigot sometimes in order to get our reactions, maybe even to keep us from being bigots. I'm never sure to what extent that was true. Moise, for instance was a Jew. Moore always claimed that Jews were inferior. I was a woman. He always pointed out that his women students were inferior. Moise and I both loved him dearly, and we knew that he supported us fantastically and did not think that we were inferior—in fact, he thought that we were super special. On the other hand, he wanted us to be very confident of ourselves in what he undoubtedly viewed as a somewhat disadvantaged position. Now then, did he play the role of a bigot to elicit a response? I have no idea.[11]

Moise managed to survive the Moore method, including Moore's Jew baiting. He became a Harvard professor, vice-president of the American Mathematical Society, and president of the Mathematical Association of America.

Although the Moore method is usually thought of as his special way of teaching graduate topology, he also taught calculus and trigonometry. Mary Ellen Rudin testified that she saw no distinction between his way of teaching, whether in trigonometry and calculus or in his most advanced graduate course. She said that in his class you didn't end up knowing much calculus because you had no training in the standard problems. So the course was not useful to a physics or engineering student. And as for the poorer students, they "dreaded and hated class for the simple reason that it was a painful experience for them. Because they had difficulty presenting things and because the things that they conjectured were often wrong, they were used as an example of how you could be wrong. It wasn't necessarily a pleasant experience for them. In fact, I think that even for the good students, it wasn't necessarily a pleasant experience altogether. You didn't enjoy seeing people fail."[12]

Another one of his students, John Green, received two Ph.D. degrees, the first with Moore and the second from Texas A. & M. in statistics. He subsequently became a principal statistician at a DuPont laboratory. Green recalls that early in the course Moore announced that "he wanted us to think about his class all day, every day, to go to bed thinking about it, to wake up in the night thinking about it, to get up the next morning thinking about it, to think about it walking to class, to think about it while we were eating. If we weren't prepared to do that he didn't want us in his class. It was also quickly evident that he meant exactly what he said."[13] Green reports that Moore would never tell the class how to prove a theorem or construct a counterexample. If the students didn't do it, it wouldn't be done. Green found that the self-reliance learned in Moore's classes was invaluable in the chemical industry. "One benefit of working with Dr. Moore," he writes, "was learning how to work with the intimidation factor. First, there is the man himself. He was a very imposing figure, he dominated his environment. We were all in awe of him on many levels. And having worked under him, no one else can intimidate me."

Moore's bitter fights with his colleagues at Texas take up many pages of Parker's biography. There were "fisticuffs" against Associate Professor Edwin Beckenbach in 1944. His most prominent mathematical colleague was Harry Schultz Vandiver, a determined and persistent attacker of Fermat's Last Theorem. (He proved it for all primes less than 2000.) He won the Cole Prize of the American Mathematical Society for his research in number theory and, like Moore, was elected to the National Academy of Sciences. Indeed, he continued producing massive amounts of research into his eighties, unlike Moore, who had long given up research in favor of producing students to become researchers. The origin of the enmity between Moore and Vandiver is unclear, but "They fell apart sometime in the late 1930s and soon they had reached the point of no contact. Their joint efforts to avoid ever having to converse with each other, let alone being in the same room if at all possible, became the talk of the university."[14]

There was even an "incident" when Moore threatened Vandiver's son Frank with a handgun. Moore owned a revolver and had been involved with gun club activities.

Changing Times

The possibility of Moore's having to admit a Black student to his course did not arise until the late 1950s, because up to then Black Texans seeking a university education were admitted only to traditionally Black state colleges.

As a result of the Supreme Court's Brown vs. Board of Education of Topeka decision in 1954, and another specifically against the University of Texas in 1956, the regents of the university declared in June 1956 that thereafter they would strive toward total desegregation. However, Robert Lee Moore never ceased defying this policy of the regents and decision of the U.S. Supreme Court.

In spite of Moore's intransigence, African-American students were admitted to the University of Texas. Walker Hunt, A. N. Stewart, and L. L. Clarkson became the first African-American students to earn Ph.D.s in mathematics from the University of Texas. Hunt reports: "I also wanted to take Robert Lee Moore's famous Foundations of Point Set Topology. However, that was not to be. The reason: I was Black! His words were, 'you are welcome to take my course but you start with a C and can only go down from there."[15]

African-American Pioneers in Mathematics

Several of the Black mathematicians who faced institutional and personal racism in their early careers interacted with Moore or his students. The first female African-American student at the University of Texas to earn a Ph.D. in mathematics was Vivienne M. Mayes. She also was denied access to a course she wished to take with Robert Lee Moore. Subsequently, Dr. Mayes became a professor at Baylor University, in Houston, Texas. In 1971 the Baylor student congress elected her outstanding faculty member of the year. (See chapter 7.)

Raymond Johnson, now a professor at the University of Maryland, earned a B.S. at the University of Texas and a Ph.D. at Rice, in Houston, Texas (the first African-American to do so). He writes, "The image of R. L. Moore in my eyes is of a mathematician who went to a topology lecture given by a student of R H Bing. The speaker was to be one we refer to as a mathematical grandson of Moore. When Moore discovered that the student was black, he walked out of the lecture."[16]

Nevertheless, Moore had at least three African-American mathematical grandsons! Interestingly, one of these, Beauregard Stubblefield, had been a student of Clarence Stephens at a TBI, Prairie View, where he earned a bachelor's degree in math-

Figure 8-3. Vivienne Malone-Mayes.
Courtesy of Mathematicians of the
African Diaspora website.

ematics in 1940. He continued at Prairie View and received a
master's in 1944.[17]

Stephens writes:

I discovered Stubblefield in my college algebra class during my
first semester as a college teacher. At the beginning of the course,
I realized that he knew the mathematics I was to teach in the
course. I told him that I was giving him a grade of "A" for the
course and I would give him a special mathematics assignment,
so that he could benefit from his studies. However, I requested
that he attend my college algebra class. At the time, students
were assigned problems to work at the blackboard. I told him
that I wanted him to help me check the solutions of students at
the blackboard. Many students in the class had finished the same

high school that Stubblefield had finished. I wanted to demonstrate that a student who had finished the same high school that many of them had could achieve well in mathematics.[18]

This was at a time before the University of Texas accepted Black students. So Stephens recommended the University of Michigan for Stubblefield's doctoral study. Three of Moore's famous White students had become professors at Michigan. Stubblefield, once enrolled at the University of Michigan, took Moore method courses from Ray Wilder and Gail Young and had many mathematical talks with Ed Moise. His thesis was supervised by Gail Young. Obviously, while Moore's students may have, at least for a time, followed Moore's style of mathematical research and teaching, they were by no means infected with his racism! Stubblefield never met Moore himself, but he heard a story about him: "One time a Black got into his class and sat for a while. R. L. Moore said "I am not going to teach any of this course until a certain person leaves my class."[19] "Stubblefield said he felt entirely comfortable with the Moore method and carried much of it into his career in mathematics, principally at Appalachian State University, and, after his retirement, as a research mathematician concentrating on number theory."[20]

Obviously, when Stubblefield and other direct or secondary descendents of Moore used the Moore method in their own teaching, it was not the unadulterated virulent original version. It was modified and adapted to circumstances, and its sheer ferocity, one might even say brutality, was toned down and humanized. Certainly Stubblefield did not hide from his students all mathematical knowledge except what they had created themselves under his control. When speaking of the Moore method used by anyone but Moore, one should think of it as an adapted Moore method.

Professor Stubblefield participates in annual meetings in Austin of an organization called the Education Advancement Foundation (EAF). This group established a Legacy of R. L. Moore Project, whose purpose is to "help advance studies of the mathematician Robert Lee Moore (1882–1972), thereby promoting the study of more effective methods of learning and teaching at all educational levels and in all subjects."[21] Dr. Nell Kroeger, who was Moore's last Ph.D. student, informed R.H. that "The EAF is deeply concerned with getting women and minorities more involved in advanced mathematics."[22]

In addition to Parker's full-length biography of Moore, there is plentiful documentation of his teaching by his students. A movie of Moore in action is available from the Mathematical Association of America. On the web you will also find an entry about him at a site called Mathematicians of the African Diaspora.

Moore had two more Black mathematical "grandsons" whom he refused to acknowledge. They were Dudley Weldon Woodard (1881–1965) and William Schieffelin Claytor (1908–1967). Both were point-set topologists, both students of John Kline, who had been Moore's first Ph.D. student. What burden did these men bear, being rejected and denied by their own mathematical grandfather? They were honored in 1999 at an exhibition, "Pioneering African-American Mathematicians," at the University of Pennsylvania.

The older of these grandsons, Woodard, earned a B.A. at Wilberforce College and a B.S. and an M.S. in mathematics at the University of Chicago. He taught at Tuskegee Institute and at Wilberforce and joined the mathematics faculty at Howard University in 1920. He was already a professor at Howard University, and also a dean, when he decided in 1927 to go beyond his master's degree and pursue a Ph.D. in mathematics. For this purpose he went to the University of Pennsylvania, in Philadelphia, which had admitted Black students since 1879, and became a student of Kline. Kline not only was a student of Moore but also continued to be a close friend and colleague of his after Moore went to Texas. After Woodard completed his doctorate in 1928, he continued teaching at Howard, and established a master's degree program in mathematics there. In 1929 he was happy to find an exceptionally promising student at Howard, William Schieffelin Claytor. He advised Claytor to go to Pennsylvania for doctoral work under John Kline, Woodard's old professor.

Claytor was a brilliant graduate student. He won the most prestigious award offered by Penn at that time, a Harrison Fellowship in Mathematics. His dissertation achieved a significant advance in the theory of "Peano continua"—a branch of point-set topology in which Kline was an expert.[23] Yet the only university job Claytor could get in 1936 was at a TBI, West Virginia State College. In 1937, on Kline's recommendation, Claytor obtained a Rosenwald Fellowship and spent a year working with Wilder and a group of topologists at Michigan. Forty-three years later, in 1980, the National Association of Mathematicians inaugurated the William W. S. Claytor Lecture Series. At that time Raymond Wilder wrote a letter about Claytor:

> Toward the end of his stay at Michigan the question of where he could get a "job" came up. We topologists concluded he should join the University of Michigan faculty. Today, I'm confident there would be no hesitancy about this on the part of the Michigan administration, but that was about thirty years ago . . . it was to no avail; the administration was simply afraid (I am sure this was the case more than racial prejudice.) I finally wrote to Oswald Veblen, head of the School of Mathematics at the Institute for Advanced Study, who quickly replied that he'd find a place for Bill at the Institute. However, when I told Bill this, he shook his head and replied, "There's never been a Black at Princeton, and I'm not going to be a guinea pig." I've always felt this was the turning point in Bill's life, and a great mistake on his part. I knew how he felt and argued with him, but he was adamant. I am sure that if he had accepted, he would have found lots of friends, at the Institute, and that his future would have been quite different.

Samuel Eilenberg wrote, "I very deeply felt the tragedy of the situation."[24] Gail Young wrote, "The two papers on which his reputation rests are brilliant. That he could not get a job in any research-oriented department was tragic."[25]

The web site, Mathematicians of the African Diaspora, reports that "Claytor did make presentations at meetings of the American Mathematical Society, but he was never allowed to stay in the hotel where the meetings were held. Instead, a home of local 'cullud' persons was found for his stay."

Claytor gave up mathematical research. According to the UPENN web site, ". . . his unfulfilled promise was a great disappointment for John R. Kline and his generation of colleagues at Penn." It's an irony of American mathematics that Princeton's loss became Howard's gain. Claytor joined the Howard University faculty in 1947, a year after Woodard's retirement, and remained there until taking early retirement in 1965. So Robert Lee Moore, by way of his mathematical son, John Kline of Pennsylvania, generated two of Howard's leading mathematics professors.

Wilder's efforts on behalf of Claytor were commendable. He probably didn't appreciate the difficulties Claytor would have confronted in 1937, just finding a place to live in Princeton. Although Princeton is in "the North," Princeton College then was a preferred place for wealthy southerners to send their sons. In Princeton in 1937, Jews and Italians were also unwelcome in the "better" sections of town. There were Black people in Princeton—domestic servants in the homes of upper-class whites. Of course, they lived in "their own" part of town. Where would a Black professor have fit in to the Princeton of 1937?

The war against Nazism ultimately led to the United States' official renunciation of racism. It was during that war that the famous probabilist Joseph Doob came from Urbana, Illinois, to visit the Institute for Advanced Study in Princeton. With him was his brilliant doctoral student David Blackwell. Princeton University as a matter of course made its libraries and seminars open to members of the Institute for Advanced Study, but in Blackwell's case they said "No." Blackwell later wrote, "I think it was the custom that members of the Institute would

Figure 8-4 (above left). David Blackwell, Berkeley statistician and probabilist. Courtesy of Archives of the Mathematisches Forschungsinstitut Oberwolfach
Figure 8-5 (above right). William Schieffelin Claytor. Courtesy of Mathematicians of the African Diaspora website

Figure 8-6 (left). Clarence Stephens and Harriette Stephens. Courtesy of Mathematicians of the African Diaspora website.

be appointed honorary members of the faculty at Princeton. When I was being considered for membership in the Institute, Princeton University objected to appointing a Black man as an honorary member of the faculty. As I understand the story, the Director of the Institute for Advanced Study just insisted and threatened, I don't know what, so Princeton withdrew its objections. Apparently there was quite a fuss over this, but I didn't hear a word about it."[26] Blackwell later was for many years chairman at Howard, and then joined the statistics department at Berkeley. He became chairman there and was elected to the National Academy of Sciences.

Clarence Stephens and the Potsdam Model

Clarence Stephens was mentioned in Beauregard Stubblefield's life story. Stephens was at Prairie View, Texas, before going on to the chair at Morgan State in Baltimore. He published two papers on nonlinear difference equations and thereafter devoted himself to teaching. His teaching philosophy was the most complete contrast to Moore's imaginable.

Stephens writes:

> When I tutored my fellow schoolmates in mathematics during my high school days, I learned that many students can learn mathematics if the learning environment is favorable. Early in my career as a college teacher, I came to the conclusion that *any college student who wanted to learn college mathematics could do so if the learning environment was favorable.* So when I was given the opportunity of leadership as the Chair of a mathematics department, I was determined to prove my conjecture. My main difficulty in proving my conjecture was that students, faculty and administration did not believe my conjecture was true. However, in my effort to prove my conjecture, I received support from each group. The students I was trying to help gave me the best support. The results we obtained at Morgan State College and at SUNY at Potsdam proved, at least to me, that my conjecture was true.[27]

At Morgan State he found that nearly all students were required to take a 6-semester-hour general mathematics course reviewing elementary and high school mathematics. This policy was based on the results of placement exams. As a consequence, very few students, even mathematics majors, could get to true college-level mathematics courses until their sophomore year. In contrast, Stephens believed that one of the best ways to prepare students for graduate study in mathematics was to place them in calculus courses as early as possible. Stephens expected that a student previously labeled "underprepared" could achieve well in such courses if the atmosphere in the department and in such courses was nurturing, and if there were role models to show that success was possible. Stephens instituted an undergraduate honors mathematics program that exposed undergraduates to first-year graduate mathematics. It drew a large percentage of Morgan State's best students to major in mathematics. Morgan State had had no student go on to obtain a Ph.D. in mathematics during the 90 years of its history before Stephens's arrival, but at least nine students who passed through its mathematics program during Stephens's years on its faculty eventually obtained that degree. One of them, Vassily Cateforis, was born in Greece. Cateforis got a Ph.D. in algebra at the University of Wisconsin and ultimately became Stephens' successor as chairman at Potsdam.

After Stephens retired from Morgan State, he taught from 1962 to 1969 at the State University of New York at Geneseo. Then he became chairman of the department of mathematics at the State University of New York at Potsdam, a small town in far northern New York State.

In March 1987, an article appeared in the *American Mathematical Monthly* with the title "A Modern Fairy Tale." It described an amazingly successful undergraduate mathematics program at a little known college, Potsdam College of the State University of New York. The author, John Poland, was in the department of mathematics and statistics of Carleton University in Ottawa, Canada. He wrote:

> Tucked away in a rural corner of North America lies a phenomenally successful undergraduate mathematics program. Picture a typical, publicly funded Arts and Science undergraduate institution of about 5,000 students, with separate departments of Mathematics and Computer Science. While the total number of undergraduates has remained relatively fixed over the past 15 years, the number of mathematics majors has doubled and doubled again and again to over 400 now in third and fourth

year. They don't offer a special curriculum. It is just a standard, traditional pure mathematics department.

More than half the freshman class elect calculus, because of the reputation of the mathematics department carried back to local high schools. And of the less than 1000 Bachelor degrees awarded, almost 20% are in mathematics. In case you are unaware, 1% of Bachelor degrees granted in North America are in mathematics. These students graduate with a confidence in their ability that convinces prospective employers to hire them, at I.B.M., General Dynamics, Bell Laboratories and so on.

Do they just lower their standards? Mathematics teachers in the university across the street [Clarkson Institute of Technology] say "no." They see no significant difference between their performance and that of their own students.

The students say the faculty members really care about them, care that each one can develop to the maximum possible level. It is simply the transforming power of love, love through encouragement, caring and the fostering of a supportive environment. By the time they enter the senior year, many can read and learn from mathematics texts and articles on their own. . . . They graduate more women in mathematics than men. They redress a lack of confidence many women feel about mathematics. In the past ten years, almost every year the top graduating student at this institution, across all programs, has been a woman in mathematics.[28]

It is rare in any mathematical publication to find words like "love" and "caring" used in this way. A welcome contrast, to the more widespread discussion of "math avoidance" and "math phobia"! Poland further writes:

What must a mathematics department do to attain this success? The faculty must love to teach, with all this means about communication, caring for students and for their development. They would teach at a pace which allows students time to struggle with the problems and resolve them, rather than primarily to cover material. . . . They would recognize that students need time to build the skill, understanding and self-confidence to handle most advanced mathematics. The faculty would encourage and reward the success of the students, bringing all or most of them to a high level of achievement (and high grades), rather than using the grade to filter the brightest and quickest students into further mathematics studies. The recipe for success at Potsdam is very simple: instill self- confidence and a sense of achievement through an open, caring environment.[29]

This atmosphere and attitude at Potsdam are largely the creation of Clarence Stephens.

Potsdam dates back to the St. Lawrence Academy, which was founded in 1816. Until 1962, its purpose was to educate students for a career in elementary teaching. It had about seven math teachers who taught mostly elementary math courses; the most advanced was introductory calculus. In 1969 the college was visited by Clarence Stephens. His talk on math and math teaching deeply impressed the math teachers at Potsdam College. They were so overwhelmed by his vision of creating a humanistic environment for learning undergraduate mathematics that they approached the administration to give him the best possible offer because "he is worth more than we can pay him.'"[30]

Stephens told Dilip Datta, a visitor to Potsdam who wrote a book about the Potsdam model:

My primary goal as Chair was to establish the most favorable conditions I could for students to learn and teachers to teach. . . .
A team of mathematics faculty members with me as a member was formed to teach students in the early (freshman and sophomore years for undergraduates—first year for graduate students) study of mathematics, "How to Read Mathematics Literature with Understanding and to Become Independent Learners." A person selected for the team was a person who, in my opinion, had a warm relation with beginning students, strong loyalty to

the department and the college. . . . I had confidence that any caring mathematics faculty member could effectively teach the students developed by the team. Also, the students who were developed by the team would help us teach other students as tutors. . . . The indicated method for developing the mathematics potential of students was as effective at SUNY Potsdam as it had been at Morgan State College.[31]

The motto of the math department at Potsdam is "Students come first." Dilip Datta writes that Stephens "insisted that every faculty office must be furnished with comfortable cushioned chairs for the students. . . . He would constantly make the teachers aware that students have other courses to study and that they need to relax over the weekends. . . . The first thing he would ask a teacher would be something like, 'Are your students enjoying math?'"[32]

Some of the things Stephens used to say are:

"Believe in your students, everyone can do mathematics."
"Know your students well—their names, what they know, their hopes and fears."
"High standards do not mean having unrealistic expectations so students feel that they have failed."
"Go fast slowly."[33]

Datta further writes, "The most important component of the humane environment of teaching and learning mathematics is the team of teachers. Without their dedicated efforts, nothing would have been possible."[34]

Stephens has received honorary doctorates of science from Lincoln University in Pennsylvania and Johnson C. Smith University, his alma mater, and doctorates of humane letters from Chicago State University and the State University of New York (SUNY). He has been honored by the governors of both Maryland and New York. He is included in the National Museum of American History of the Smithsonian Institution. The Seaway section of the Mathematical Association of America renamed its annual Distinguished Teaching Award the Clarence Stephens Award. That means that his name is announced every year at the section's award ceremony.

In 2003 the Mathematical Association of America gave him its "Gung-Hu award" (officially called the Gung and Hu Award) for outstanding service to mathematics. The citation quotes Stephens' description of his teaching methods, and comments, "Though SUNY Potsdam is a relatively small regional state college with a total enrollment of just over 4,000 students during Stephens' time there, in 1985 the college graduated 184 mathematics majors, the third largest number of any institution in the U.S. that year (exceeded only by two University of California campuses). This represented about a quarter of the degrees given by SUNY Potsdam that year, and over 40% of the institution's honor students were mathematics majors."

More important than a teacher's personal awards is the influence of his teaching methods. Robert Lee Moore certainly had a considerable influence in the United States. From 1950 onward, five of Moore's doctoral students, and a sixth who studied with him, served as presidents of the Mathematical Association of America. The fact that for decades the MAA was led by a student of R. L. Moore must have significantly affected college education. Moore even affected elementary and secondary education, through his student Ed Moise and through Ed Begle, a student of Ray Wilder who was thus a second-generation Moore student. Moise and Begle were important leaders of the School Mathematics Study Group (SMSG), which for a while managed to push sets and axioms into elementary school math teaching. At the university level, the Moore method, or rather the modified Moore method, survives as a teaching method. It is unusual and nonstandard but respected as a proven way to educate future research mathematicians.

I (R.H.) visited Potsdam in 2002 and found the Stephens spirit and attitude prevailing under Cateforis' chairmanship. Stephens had retired in 1987, the year John Poland's article, "A Modern

Fairy Tale," appeared in the *American Mathematical Monthly*. Stephens' efforts did not end with retirement. In response to a question, he wrote:

> I received invitations to visit colleges and universities in Canada and throughout the United States in order to discuss the Potsdam mathematics program. I visited colleges and universities in the East, Midwest, West, and South. I visited almost all of the California State Universities and received invitations from a few that I did not have the time to visit. I made several return visits. After 4 years of making these visits, I stopped accepting invitations. In California and Georgia, I was offered faculty positions in order to help establish similar programs. From my experiences at Morgan and Potsdam I knew that it was difficult to establish favorable academic environments, so that any college student who desired to do so could learn and enjoy learning mathematics. To establish a successful program depended on *creative thinking, time, and place*. Hence, I did not accept any faculty position offered to me.[35]

And what about the Stephens method, the Potsdam model? When I visited Potsdam, I asked the faculty whether other math departments emulated their method. They answered, "People come here and watch. Then they say, 'This is great, but we could never do it on our campus.'"

However, when I asked Clarence Stephens whether his model had been adopted elsewhere, he named several prestigious schools. Cal Tech—California Institute of Technology! And Princeton! And Dartmouth! And two other selective private schools of somewhat lesser fame: Spelman College in Atlanta (a traditionally Black women's institution) and Harvey Mudd College in Claremont, California (a high-prestige engineering-oriented school, one of the well-known Claremont Colleges). But it has not been adopted by one state university or community college, where the great majority of American college students are found.

This is disappointing, but understandable. At Potsdam, Stephens had to persist for years in order to win his whole faculty over to his point of view: "Under favorable conditions, any college student interested in learning mathematics can be successful!"

And what are favorable conditions?

"Students come first."
"Give a student all the time he or she needs, to absorb the material."
"Have complete confidence that every student can be successful."

At Potsdam, the faculty is not under constant strong pressure to publish as much as possible. And mathematics is there for any student who is interested in learning it—not only future scientists and engineers. Many of the math majors at Potsdam become mathematics teachers, but others go to work in business or other nonacademic careers. Yet many of them find that their mathematics education is useful in their life.

But the prevalent view in U.S. college education is different. "You major in the subject that will ultimately land you a good job." If you take calculus but you aren't a math or science major, it's in order to get into business school (or medical school or architecture school).

It would be quite a project to convert an ordinary state university in the United States to the Stephens philosophy. Faculty are to be rewarded for putting students first? And any student who is interested in mathematics will succeed? And each student of mathematics feels loved?

However, here is one consideration that might move faculty and administrators to pay attention to the Potsdam model. Many U.S. universities are struggling to increase the success rate of minority students in mathematics. (In chapter 10 we report on Uri Treisman's successes at Berkeley and Austin.) At any school where the Potsdam model is implemented, one could confidently expect great improvements in the success rates of all students—both minority and majority.

Conclusions

The stories of Clarence Stephens and Robert Lee Moore embody two different, opposed strains in American education: the egalitarian versus the elitist; the cooperative versus the competitive; the heritage of the Declaration of Independence versus the heritage of the Confederate States of America. Their stories reveal that while mathematical life may at times appear to be an ivory tower where we escape from social conflicts, it may at times be a maelstrom where social currents clash. Moore's barring African-American students from his class was part of a long legacy of racism in the United States, especially in the "slave states" that sought to secede from the Union.

Stephens, on the other hand, grew up in African-American communities that took pride in traditional Black institutions. The beliefs and values of his subculture included mutual support. For him, teaching was more than the transmission of specialized knowledge; it meant full acceptance of each individual in his or her efforts to create a better life.

The southern segregationist subculture to which Moore owed allegiance is now considered defeated and disreputable. This was demonstrated by the election of an African-American President of the United States. But full integration of previously excluded groups is still to be achieved. It requires more than mere legal equality, it demands transformative teaching methods.

Bibliography

Albers, D. J. (1990). *More mathematical people.* Boston: Harcourt Brace Jovanovich.

Albers, D. J., & Alexanderson, G. L. (Eds.) (1985). *Mathematical people: Profiles and interviews.* Boston: Birkhäuser.

Datta, D. K. (1993). *Math education at its best: The Potsdam model.* Framingham, Mass.: Center for Teaching/Learning of Mathematics.

Donaldson, J. A. (1989). Black Americans in mathematics. In Peter Duren (Ed.): *A century of mathematics in America,* part III. Providence, R.I.: American Mathematical Society.

Einstein, A. Quoted in Holton, G. (1973). *Thematic origins of scientific thought: Kepler to Einstein.* Cambridge, Mass.: Harvard University Press.

Lewis, A. C. (1976). George Bruce Halsted and the development of American mathematics. In J. D. Tarwater (Ed.). *Men and institutions in American mathematics.* Lubbock, Tex.: Texas Technical University Press.

Mathematicians of the African Diaspora web site. Retrieved April 5, 2007, from http://www.math.buffalo.edu/mad/

Megginson, R. E. (2003). Yueh-Gin Gung and Dr. Charles Y. Hu award to Clarence F. Stephens for distinguished service to mathematics. *American Mathematical Monthly 110*(3), 177–180.

Parker, J. (2004). *R. L. Moore: Mathematician and teacher.* Washington, D.C.: Mathematical Association of America.

Poland, J. (1987). A modern fairy tale? *American Mathematical Monthly 94*(3), 291–295.

Stephens, R. (2006). Personal communication.

Stephens, R., web site. Retrieved March 15, 2007, from http://www.mathsci.appstate.edu/~sjg/womenandmonoritiesinmath/student/stephens/sephen

Wilder, R. L. (1982). The mathematical work of R. L. Moore: Its background, nature and influence. *Archive for the History of Exact Sciences 26,* 73–97.

+9+

Loving and Hating School Mathematics

Why do so many school children and adults find mathematics intimidating and consider themselves hopelessly incapable of learning it? How do educators address this challenge?

Mathematical life is an immersion in a world of endlessly varied forms and relations. The mathematician is challenged and tempted to commit all her energy and enthusiasm to learn and to understand. Mathematical thinking is also enjoyed by people working puzzles, playing chess, or doing recreational problem solving. Engagement with and enjoyment of mathematics is the primary topic of this book.

But there's also another thing called mathematics. It's the thing people are talking about when they say:

"I hate math! I couldn't learn it, and I can't teach it!"
"I'm bad at math. It's always been my least-liked subject."
"I hated math in school . . . and my feelings haven't changed since."

These comments about *school mathematics* come from graduate students, who were actually preparing to become elementary school teachers by taking a seminar on math instruction. "The students were nearly evenly divided between those who liked and those who disliked math. In nearly all the cases, a correlation existed between attitude and success."[1] What emotions about mathematics will children absorb from such teachers?

There is a large literature on negative attitudes toward mathematics. The best known is Sheila Tobias' book on "math phobia."[2] The issue of math phobia as experienced by individual students has been extensively explored. We wish to address related issues. One of these is the conflict between society's need for mathematical engineers and scientists and the difficulties of many individual students who are not preparing for such careers. How can these two concerns be reconciled? Does everybody need to be a math expert in order to adequately respond to the demands of the information age?

A related question is, How much mathematics is actually needed for a career such as medicine? Today, studying mathematics isn't just an option a student can choose according to her interest. It's compulsory. Proficiency in algebra, and in some instances calculus, is considered essential for many professions. Math serves as a filter to screen applicants for college and for professional schools. How realistic is this requirement? Should your admission to medical school be contingent on your grade in calculus? This is another issue we explore in this chapter.

Reluctant Learners

It's a common observation that a large proportion of students are profoundly alienated from mathematics. And this problem of avoiding and rejecting math isn't disappearing. Newspaper headlines bemoan the poor standing of the United States in international comparisons of mathematical achievement and blame schools, television, and parents' unavailability when their children need help with homework. With the passage of the federal No Child Left Behind Act, the pressure has become more intense. Schools that don't test within the mandated range are penalized. Children whose performance lags behind the standards are stigmatized. These punishments contribute to math phobia. Popular surveys as well as more systematic studies find that mathematics is the school subject that provokes the strongest reactions, both negative and positive.

The following article had the headline, "Hate mathematics? You are not alone."

> People in this country have a love-hate relationship with math, a favorite school subject for some but just a bad memory for many others, especially women. In an AP-AOL News poll as students head back to school almost four in 10 adults surveyed said they hated math in school, a widespread disdain that complicates efforts today to catch up with Asian and European students. Twice as many people said they hated math as said that about any other subject. Some people, like Stewart Fletcher, a homemaker from Suwannee, Georgia, are fairly good at math but never learned to like it. "It was cold and calculating," she said. "There was no gray. It was black and white." Still, many people, about a quarter of the population, said math was their favorite school subject.[3]

Notice that while 40 percent hate it, there are 25 percent who like it better than any other subject! A lot of people do hate math, but a lot of other people love a chance to challenge their brains with math problems. There are many people who enjoy doing something where there is just one right answer and every other answer is wrong.

In his *Apology*, Hardy wrote, "The fact is that there are few more 'popular' subjects than mathematics . . . there are probably more people really interested in mathematics than in music. . . . There are masses of chess players in every civilized country. . . . Chess problems are the hymn-tunes of mathematics. We may learn the same lesson, at a lower level but for a wider public, from bridge, or descending further, from the puzzle columns of the popular newspapers. Nearly all their immense popularity is a tribute to the drawing power of rudimentary mathematics . . . nothing else has quite the kick of mathematics.[4]

A puzzle that was a great craze over a hundred years ago is still popular today. The Fifteen Puzzle is about sliding numbered blocks inside a square frame. Back in the spring of 1880, the *New York Times* wrote: "No pestilence has ever visited this or any other country which has spread with the awful celerity of what is called the 'Fifteen Puzzle.' It has spread over the entire country. Nothing arrests it. It now threatens our free institutions, inasmuch as from every town and hamlet there is coming up a cry for a 'strong man who will stamp out this terrible puzzle at any cost of Constitution or freedom.'"[5]

Today, Su Do Ku, a puzzle about arranging numbers in a square box, is millions of people's favorite game. In 1997 a retired Hong Kong judge, Wayne Gould, saw a partly completed puzzle in a Japanese bookshop. Over 6 years he developed a computer program to produce puzzles quickly. He promoted Su Do Ku to the *Times* in Britain, which launched it in 2004. By April and May 2005 the puzzle was part of several other national British newspapers. The world's first live TV sudoku show, "Sudoku Live," was broadcast on July 1, 2005. Nine teams of nine players (with one celebrity on each team) representing geographical regions competed to solve a problem. Addiction to math games and puzzles by one population coexists with the rejection of anything mathematical by many others. It's likely that among those who say they don't like math, or even hate it, are many who enjoy mathematical challenges in the form of games and puzzles.

When we hear somebody say he or she doesn't like math or avoids math, we ask, "When did it start?" The answer is either "4th grade" or "6th grade" or "8th grade." Recently, at dinner, our friend Claire answered, "6th grade." She elaborated: "The teacher just called on boys, he didn't think girls could really do math. And also, my friends would get on me, for being 'too smart.'"

I said, "So your teacher thought you were too dumb, and your friends thought you were too smart?"

"Yes, that's what happened."

For some who dislike mathematics, it starts with fractions. (In fact, most adults in the United States have serious trouble adding 1/3 + 1/4.) For many others, it's algebra, working with x and y. And for others, who thought they did all right in arithmetic and algebra, it's their college calculus course that convinces them they're "not a math type" or are even "bad at math." *People aren't born disliking math. They learn to dislike it in school.*

The first meeting with algebra in middle school—grades 6 through 8, ages 12 to 14—seems to be critical for many students. Unfortunately, this stage has received less study by educational researchers than have the early childhood and beginning primary school years. As Kristin Umland[6] has pointed out, this stage of school is where the transition must be made from the "premathematical" to the "fully mathematical." Roughly speaking, from the concrete to the abstract, or, as Bertrand Russell put it, from thinking about a particular thing to thinking about an unspecified member of a whole class of things. This leap is easy for some, but more difficult for others. We teachers need to understand better how to help children overcome this difficulty.

The current emphasis is on test scores, punishing individuals or schools that fall short of a prescribed level of tested mastery. To meet international competition, we increased the number of required algebra and trigonometry courses in high school.

Some have questioned this approach. In the *Washington Post*, columnist Richard Cohen criticized a new requirement of a year of algebra and trigonometry for graduation. He argued that this contributes to a higher dropout rate. Cohen recalled his own terror: "There are those of us who know the sweat, the pain, the trembling, cold fear that comes from the teacher casting an eye in your direction and calling you to the blackboard. It is like being summoned to your own execution"[7]

In an earlier article, Colman McCarthy, the founder of the Center for Teaching Peace, made a similar point: "Too many of us were forced to take algebra when the time and energy could have been devoted to subjects that were truly beneficial individually and nationally. Algebra isn't essential to much of anything. Once adding, subtracting, multiplying and dividing are mastered—by eighth grade usually—why insist on more? Algebra . . . is a language, a way of symbolic communication that a few people find fascinating and practical, while most don't. Would millions of high school students trudge into their algebra classes if there weren't a gate through which they were forced to pass to enter college?"[8]

He further argued: "The world is crying out for peacemakers. We are not teaching the kids how to be the essential thing. We have conflicts all our lives."[9] In response, some people would argue that mathematics that is connected to daily life also addresses issues of conflict.

To these two contemporary voices, we can add one more from a century ago. The famous philosopher Bertrand Russell was coauthor, with Alfred North Whitehead, of the monumental *Principia Mathematica,* an epochal work that attempted to reduce all mathematics to symbolic expressions from formal logic. Yet this master of rigorous formal mathematics had misgivings about school algebra. In 1902 he wrote:

In the beginning of algebra, even the most intelligent child finds, as a rule, very great difficulty. The use of letters is a mystery, which seems to have no purpose except mystification. It's almost impossible, at first, not to think that every letter stands for some particular number, if only the teacher would reveal what number it stands for. The fact is, that in algebra the mind is first taught to consider general truths, truths which are not asserted to hold only of this or that particular thing but of any one of a whole group of things. . . . Usually the method that has been adopted in arithmetic is continued: rules are set forth, with no adequate explanation of their grounds; the pupil learns to use the rules blindly, and presently, when he is able to obtain the answer that the teacher desires, he feels that he has mastered the difficulties of the subject. But of inner comprehension of the processes employed he has probably acquired almost nothing.[10]

Does it have to be this way? Some educators are trying different ways to teach mathematics in school, in the United States, as well as in many other countries, as described in the following section.

School Mathematics and Everyday Mathematics

Imparting basic mathematical knowledge to children and young people who are uninterested or fearful is a serious undertaking. They must learn to add, subtract, multiply, and divide fractions and also acquire some basic geometry and algebra. These are challenging tasks. They require strong teaching, with an emphasis on conceptual understanding, and most important, with connections to activities that are relevant to children's lives.

Children enter school with a variety of experiences with shapes, categories of objects, estimation of areas, and some counting. According to the Swiss psychologist Jean Piaget, young school-age children are in the process of mastering conservation of quantity, seriating, and equivalence of corresponding sets, based on both visual alignments and counting. But these notions are acquired slowly. Many 5-year-olds can recite numbers but have not yet mastered the concept that counting means amount, something that remains the same even if the objects being counted are rearranged.

The context in which such mastery takes place varies greatly. Walkerdine (1997) suggests that even simple conceptual pairs such as "more" and "less" need to be rethought. Many children (particularly those raised in poverty) hear "more" as paired, not with "less," but with "no more."[11] The operations of addition, multiplication, and subtraction are embedded differently in different languages. In French, 90 is *quatre-vingt-dix* (four twenties and ten); its name uses both multiplication and addition. In the West African language Yoruba, 35 is named as "five from two twenties," using multiplication and subtraction.[12] The Masai of Kenya signal the number 8 by raising four fingers of the right hand and waving them twice. [13]

Some children enter the world of mathematical patterns more comfortably through visual experiences than through language. The Hopi of the American Southwest grow 24 varieties of corn. Children start acquiring basic mathematical concepts by helping to sort the corn according to color and size. The challenge is to build on these concepts in the classroom.[14]

Informal knowledge of geometric concepts is contained in traditional crafts and construction. The Mozambican mathematician Paulus Gerdes describes how in some African communities bamboo sticks are joined with ropes and shaped into rectangles to make components for a house. Mathematics teachers can use this familiar activity of artisans and house builders to introduce geometry to young learners.[15]

Young street vendors in Brazil accurately perform complex mental calculations, way beyond what they can achieve by school methods.[16] Rather than multiply, they "perform successive additions of the price of one item, as many times as the number of items to be sold."[17] They use concrete referents and operations with which they are very familiar. "At the same time, while their everyday mathematics provides them the anchoring of specificity, it limits their flexibility."[18] Because they use repeated adding rather than multiplying, the street kids don't learn the commutative law of multiplication. On the other hand, children in school who do know the laws of arithmetic may make careless mistakes; their errors will not cost them money.

Educators are looking for ways to connect everyday mathematics and school mathematics by introducing contexts that are meaningful to children. Jere Confrey, a well-known Piagetian mathematics educator, uses the metaphor of "splitting," which includes sharing and mixing activities that are part of children's daily lives. For example, mixing concentrated liquids to make lemonade is a way to teach proportions. As the children shift from one kind of drink to another, and change the quantities to be produced, they learn about ratios. They work in groups and then discuss their different approaches, providing insights to the teacher-researcher listening to the students' voices. "It seems clear that . . . children can operate intelligently with ratios, especially if they are provided access to appropriate representations (data tables, ratio boxes and two dimensional plane) within interesting and familiar contexts."[19]

Research on everyday mathematics is carried on in varied contexts, such as shopping, farming, sewing, and marketing. The anthropologist Jean Lave believes that knowledge about human problem solving is best pursued in "the experienced, lived-in world as the site and source of further investigations of cognitive activity."[20] She studied the use of arithmetic by adults who focused on best-buy strategies while shopping for groceries. The shoppers compared prices, occasionally using hand calculators. They also were concerned with other issues, such as storage space and trying out new recipes. Participants sometimes used direct manipulation. One dieter needed to make a serving of three-quarters as much cottage cheese as the two-thirds of a cup allowed by her diet. If she had been in a classroom, she would have been expected to multiply 3/4 times 2/3 and cancel the 3's to get the answer, 2/4 = 1/2. Instead, she solved the problem physically. She "filled a measuring cup two-thirds full of cottage cheese, dumped it out on a cutting board, patted it in a circle, marked a cross on it, scooped away one quadrant, and served the rest."[21] Thus, algorithms taught in school are not always transferred directly to everyday uses. But skills that are decontextualized in school can become alive and useful when applied to life experiences.

Math Reform

Studies like Confrey's and Lave's lead to new approaches to math instruction, connecting learning to real life experiences. There have been many efforts to reform math education. Some of them focus primarily on cognitive approaches, developing children's number sense, mental mathematics, and understanding of patterns. One of these was begun by the famous Dutch mathematician Hans Freudenthal. Freudenthal was born to a Jewish family in Luckenwalde, Germany, in 1905. He received a Ph.D. in 1931 at the University of Berlin, under Heinz Hopf. He was then already in Amsterdam, having been invited there in 1930 to become Brouwer's assistant. He soon mastered the Dutch language and even won first prize in 1944 for a novel! At that time, Holland was under Nazi occupation, and Freudenthal and his family were in hiding. A friend played the dangerous role of prizewinner at interviews, dinners, and speeches. But the badly needed prize money did reach Freudenthal and was a great help in surviving the last year of the war. When Holland was liberated, Freudenthal was appointed "professor of pure and applied mathematics and the foundations of mathematics" at the University of Utrecht. He became famous for contributions to topology and algebra, especially on the characters of the semisimple Lie groups.[22] He also was a prolific contributor to the history of mathematics.

In 1971 he founded the Institute for the Development of Mathematical Education in Utrecht. (In September 1991, after his death, it was renamed the Freudenthal Institute.) He is credited with "single-handedly" saving the Netherlands from following the worldwide "new math" trend in the 1970s. His institute developed "realistic mathematics education," which is based on problems taken from day-to-day experience. He taught that you learn mathematics best by reinventing it. He died quietly in October 1990, while sitting on a park bench, where he was found by playing children.

Freudenthal's approach is being developed in New York by Catherine Fosnot and her collaborators. They describe their attitude to standardized calculating procedures, or "algorithms": "Exploring them, figuring out why they work, may deepen children's thinking. . . . They should not be the primary goal of computation instruction, however. . . . Children who learn to think, rather than apply the same procedures by rote regardless of the numbers, will be empowered."[23]

Everyday Mathematics is a popular program developed at the University of Chicago that uses similar constructivist principles. It emphasizes real-world situations like Confrey's lemonade example and combines activities with the whole class, small groups, partners, and individuals. There are many opportunities for students to discuss their insights with each other and to compare their strategies. Pupils are encouraged to use calculators selectively, without making them into simple crutches.

Most reform efforts are based on constructivist theories that use counting as the basis of their instruction in arithmetic. A somewhat different approach is proposed by the Russian psychologist Vladimir V. Davydov. He was deeply influenced by the ideas of Lev S. Vygotsky who has been called the "Mozart of psychology." A cultural historical theorist, Vygotsky was familiar with Piaget's work, and the two theories do have some commonalities. One of their differences is the emphasis on "scientific concepts". Vygotsky presents the notion that teachers should introduce broad concepts that cannot be acquired solely through everyday experience. They require carefully planned teaching.

In contrast with Piaget's constructivist approach and its reliance on counting, Davydov emphasizes measurement as the basis of mathematical generalization. Both programs begin with concrete experiences such as the comparison of weights, areas, and children's heights. However, these actions are represented schematically in the Russian approach. With increasingly complex problems, children invent new forms of representation. One of the advantages of this approach is that it provides a means for reconstructing a problem. By inventing multiple schemata, children overcome the challenge of fractions and square roots.

The American educator Jean Schmittau[24] replicated Davydov's program in a northeastern school and found that children starting first grade used inventive methods in comparing quantities. They were able to engage in theoretical generalizations and were not troubled, as many of their American peers are, by multiplying ratios. Some of Davydov's methods parallel historical inventions in mathematics. To date, these methods are not well known in the West, but a recent article about the program in MAA On-Line[25] may result in wider application.

Some reform programs include parents as well. One of these is the Family Math group of activities developed at Lawrence Berkeley National Laboratory in California. It, too, emphasizes manipulation, games, and everyday experience. "There is a growing sense that one of the goals of school mathematics is to help students make sense of both standard and nonstandard algorithms."[26]

The primary emphasis of reform programs is making mathematics *cognitively* accessible to learners. But reform can go beyond intellectual challenge; it can also include the *emotional* aspects of learning mathematics. This is important in overcoming the severe underrepresentation of minorities in mathematically oriented careers. "Blacks make up perhaps 15 percent of this country's population, yet in 1995 they earned 1.8 percent of the Ph.D.'s in computer science, 2.1 percent of those in engineering, 1.5 percent in the physical sciences, and 0.6 percent in mathematics".[27] Of the programs attempting to change this situation, the most influential is Robert Moses' Algebra Project.

Moses was an outstanding leader in the struggle for voting rights in Mississippi during the 1960s. Now he is bringing proficiency in algebra to middle-school students in America's Black communities. "Our aim is to change the situation that currently exists, where large percentages of minority students who go through a high school and get admitted to a college have to take remedial math in order to get to the place where they can even get college credit mathematics courses."[28] He aims to make algebra enjoyable, by linking it to spatial concepts including travel, and by using multiple representations of mathematical notions. He is committed to giving young people a voice, to "involve the youth in all aspects of decision making".[29] He views competence in mathematics, taught and acquired in middle school,

Figure 9-1. Bob Moses during his youth as a civil rights leader. © 1978 George Ballis/Take Stock

as one of the civil rights of Black students. His program empha-sizes community participation, peer instruction, and develop-ment of a stronger self-concept by learners. It is innovative in the way that mathematics is taught, and its particular strength is that it mobilizes all of a learner's resources, including his or her emotions. It provides the balance of thought and emotion that so many mathematicians see as central to their enjoyment of their profession.

The participants in the Algebra Project view themselves not only as individual learners struggling with difficult ideas but also as members of a larger community. Teachers and coordina-tors share productive ideas, older students tutor younger ones, and all of them value the worth and potential of each individual learner. Using trips and directions as a key metaphor, they move from concrete experiences to increasingly more sophisticated algebraic expressions. Students receive tutoring from college student volunteers, older graduates of the Algebra Project, and peers helping each other. "The Mississippi kids could be taught in part by their own generation and learn more easily than the older generation being taught in the same workshop."[30] "There is a way that young people reach young people, are able to touch each other that in my view is central to the future shape of the Algebra Project."[31] Young kids like to hang out with older kids. "And this hanging out didn't have to be on street corners." Moses and Cobb include many quotes from kids describing their experiences. For instance, Heather in Jackson, Mississippi, said, "My friends question me a lot about what I do. I don't think they understand when I tell them that I leave school and go to work in the math lab. They say, 'What do you mean you are going to work? That's not working. You are just going over there and [play] with those computers.' Working is McDonalds or Jitney Jungle, to them. They feel like I'm just learning, you know. And most people don't put work and learning together."[32]

Young tutors like Heather are modeling what they themselves have learned. They have changed by being listened to, encouraged, and given responsibility. This is the emotional aspect that makes the project so effective. Even while being forced to slip under the rigid requirements of the present federal legislation affecting schools, the Algebra Project is contributing to greater mastery, self confidence, and self-respect in students who might otherwise have turned off and dropped out of school. But as Moses warns, "A network of tradition for this, involving teachers, students, schools, and community, isn't established in one fell swoop. You go around it and around it, and you keep going around it and deepening it. You keep returning to it until all of the implications of what you are doing become clear and sink in."[33]

For college students, the positive impact of group interaction was revealed in a famous study that Uri Treisman conducted with Rose Asera at Berkeley in the late 1970s and early 1980s. They were trying to understand why many Black students who had performed well in high school dropped out of the calculus sequence after arriving at Berkeley. They noticed that Chinese students did much better than Black students. Treisman discovered the key difference: while the Black students studied alone, the Chinese students worked together in group sessions. They asked each other many questions, critiqued each other's approaches, and helped each other with homework. "They might make a meal together and then sit and eat and go over the homework assignment. They would check each other's answers and each other's English. . . . A cousin or older student would come in and test them. They would regularly work problems from old exams, which are kept in the library."[34]

Based on their findings of this contrast with Chinese students, Treisman and his colleagues developed a new kind of intervention to assist minority students. They organized workshop communities where students met in addition to regular classes. These communities were not labeled "remedial classes" but rather were special opportunities for hard-working students. They were given challenging problem sets and were approached as members of an honors program, rather than as needing remediation. The program was an outstanding success, and over the last couple of decades it has been adopted at a variety of institutions.

The Treisman model bears some resemblance to the informal communities of research mathematicians, who enjoy exploring new problems and solutions with each other. Earlier in this book, we have described the crucial, supportive role of the Anonymous Group in Budapest, which served as an important setting for Paul Erdős, the best known collaborator in 20th century mathematics. In *Indiscrete Thoughts*, Gian-Carlo Rota writes of the Massachusetts Institute of Technology (MIT) commons room, where "at frequent intervals during the day, one could find Paul Cohen, Eli Stein, and later Gene Rodemich excitedly engaged in aggressive problem solving sessions and other mutual challenges to their mathematical knowledge and competence."[35] In recognizing this reliance on conversations, advice, and arguments in parks, in cafes, or on the streets of Princeton and Göttingen, we challenge the notion that mathematics is the creation of isolated individuals. It is a socially created human endeavor for mature mathematicians as well as for students.

The reform movement has met strong opposition. The antireform people seemed to focus on textbooks, and wanted to move back to basics. Their position was publicized in the mass media. For example, this article appeared in the *New York Times*:

> In Seattle, Gov. Chris Gregoire has asked the state Board of Education to develop new math standards by the end of next year to bring teaching in line with international competition. . . . Grass-roots groups in many cities are agitating for a return to basics. . . . Schools in New York City use a reform math curriculum, Everyday Mathematics, but some parents there, too, would like to see that changed. . . . A spokesman for the New York City Department of Education said that Everyday Mathematics covered both reform and traditional approaches, emphasizing knowledge of basic algorithms along with conceptual understanding. He added that research gathered recently by the federal Department of Education had found the program to be one of the few in the country for which there was evidence of positive effects on student math achievement. . . . The [antireform] frenzy has been prompted in part by the growing awareness that, at a time of increasing globalization, the math skills of children in the United States simply do not measure up: American eighth-graders lag far behind those from Singapore, South Korea, Hong Kong, Taiwan, Japan and elsewhere on the Trends in International Mathematics and Science Study, an international test. Many parents and teachers remain committed to the goals of reform math, having children understand what they are doing rather than simply memorizing and parroting answers.

Traditional math instruction did not work for most students, say reform math proponents like Virginia Warfield, a professor at the University of Washington. "It produces people who hate math, who can't connect the math they are doing with anything in their lives," Dr. Warfield said. "That's why we have so many parents who see their children having trouble with math and say, 'Honey, don't worry. I never could do math either.'"[36]

The argument of the traditionalists concentrates on test scores. They claim that it's the reform textbooks and curricula that result in the poor ranking of the United States in international evaluations. But it was the traditional curriculum and teaching that gave us adults who can't add fractions, and 40 percent of whom say they "hate" math.

The critics of reform favor the Singapore textbook. Students in Singapore score highest on the international comparison of math test scores. The Singapore text doesn't waste much time and space on needless motivation and explanation. It provides explicit directions and plenty of exercises, both easy and hard. It fits with the goals of the critics (see the web site Mathematically Correct, for example) who support the clear and direct teaching of algorithms and who stress competence in basic skills. What is good about this curriculum needs to be understood and explained, and ultimately integrated into curricula with more varied problem types and broader learning goals.

There can be many reasons for the low ranking the United States has received in international tests. Teachers in this country receive little respect, limited time for preparation, poor pay, and few opportunities to work with each other to develop stronger programs. In poor communities schools are badly neglected and convey the message, "You do not matter, and we do not expect you to succeed." In some other cultures, learning is held in higher respect than wealth. If we want to strengthen the competitive position of American learners, we need more than curricular reform. We need to reform our economics, politics, and culture.

In the meantime, while some mathematical skill is needed for everyday life, a great deal of it is needed for science and technology. Computers and their software are the central nervous system of our society. On the one hand, the development, manufacture, and use of computers absolutely require workers with advanced mathematical training, but on the other hand, the universal use of computers at home and at work makes even elementary arithmetic unnecessary for nearly everybody else. These two opposite effects of the computer revolution put math education under acute tension. There is a wrenching strain between opposing pressures: a continuing demand for enough sophisticated math specialists, with a shrinking need for traditional math skills in the general population. Math reform has to strengthen the training of those who want and need advanced mathematical skill without alienating the large population that thinks they don't need it.

In addressing this tension, we propose an approach that is developmental rather than oppositional. A long-term solution could realize some aspects of what each is advocating. Learning mathematics requires sufficient drill and practice, and it also requires challenging ideas. But it doesn't require unchanging universal standards or compulsive reliance on test scores.

A Different Perspective

Basic arithmetic is necessary to survive in a postindustrialized society. Studying it should continue to be required, but not in such a manner that students remember it with antagonism and loathing. Is the solution to make school math more like *real math*, the math enjoyed by people who love math? The more it is taught with the goal of understanding, with willingness to engage in playful exploration, and with connections to job-related mathematics in adulthood, the more chance it has to succeed.

Why do so many people at some point in elementary or middle school "hit a wall" and give up on math? On one hand, there is a continuing demand for all children to master the basics: become skilful at timed tests of arithmetic, geometry, and algebra computations. It is demanded that their test scores rise to compare better with children in Bulgaria or Singapore.

On the other hand, there is the plain fact that once out of school, hardly anybody ever has to solve a quadratic equation or prove a geometry theorem. Yes, they have to do it to get into college or to get into many graduate or professional schools. But for many, once school is over, much of mathematical learning is forgotten.

Politicians and spokespersons for academia regularly issue statements decrying the low mathematical competence of our children. (The mathematical competence of adults is seldom mentioned as a problem.) It is argued that you have to be good at math if you want to make a good living and that a more mathematically competent workforce is needed in order for our country to compete in the world economy.

But do you know a doctor, lawyer, or businessperson who uses calculus, or even an algebraic equation or a theorem from geometry? When our country's economic troubles are discussed in the business section of the *New York Times* (as opposed to the education supplement), the mathematical competence of the American population is never an issue. The U.S. steel industry collapsed because the costs of production and modernization of facilities are much higher here than in Brazil or China. General Motors and Ford are collapsing because of their high pension obligations and their weak competition in price and design against Japanese manufacturers—not because of poor algebra skills among the members of the United Auto Workers union in Detroit. America's "back-office" computing is being outsourced to India because Indian computer techs work much cheaper than Americans—not because American computer techs know less arithmetic, algebra, or calculus.

In 1997 Underwood Dudley, the incoming editor of the *College Mathematics Journal*, derided the claim (in "Everybody Counts," a document of the National Research Council) that "over 75% of all jobs require proficiency in simple algebra and geometry, either as a prerequisite to a training program or as part of a licensure examination."[37] Dudley commented: ". . . this is silly. Just look at the next eight workers that you see and ask yourself if at least six of them require proficiency in algebra to do their jobs. . . . Almost all jobs, I counter-assert, require no knowledge of algebra and geometry at all. You need none to be President of the United States, nor to be a clerk at Walmart, nor to be a professor of philosophy. . . . You might think that engineers, of all people, would need and use calculus, but this seems not to be so."[38] Dudley quotes Robert S. Pearson: "My work has brought me into contact with thousands of engineers, but at this moment I cannot recall, on average more than three out of ten who were well versed enough in calculus and ordinary differential equations to use either in their daily work."[39]

Professor Dudley concluded, "It is time to stop claiming that mathematics is necessary for jobs. It is time to stop asserting that students must master algebra to be able to solve problems that rise every day, at home or at work. It is time to stop telling students that the main reason they should learn mathematics is that it has applications. We should not tell our students lies. They will find us out, sooner or later."[40] (Dudley has been teaching calculus at DePauw University for almost 40 years.)

The inability of most Americans to add 1/4 and 1/3 correctly is embarrassing because they should have learned that in the 4th or 5th grade. If they ever need that sum, their calculator will give an answer close enough for practical purposes. But the principle involved in adding fractions could be taught more effectively than it is now. It was a real problem to some inmates at the New Mexico State Penitentiary Minimum Security Facility, where R. Hersh volunteered as a tutor for 5 years. When they got out of prison they would need a high school diploma in order to be employable, and they needed to add 1/3 + 1/4 in order to get a high school diploma.

It's not that these adults hadn't been drilled on adding fractions. They had been drilled, and drilled, and drilled again. The current No Child Left Behind Act intensifies testing and drilling by penalizing schools that don't get the demanded test score results. As a result, math education in the United States nowadays is dominated by teaching to the test.

No wonder some people hate this! Suppose you couldn't get a high school diploma without being able to sing, on key, the "Star-Spangled Banner" and half a dozen other "basic" numbers. We would produce a lot of people who hate to sing.

Fortunately, Hersh's junior high school music teacher separated her class into "high boys," "high girls," "low boys," "low girls," and "listeners." I (R.H.) was a happy listener. But my physical education instructor demanded over and over that I learn to climb a rope. That repeated humiliation of course intensified my dislike of physical education.

Many teachers, math educators, and mathematicians are trying to humanize school mathematics. They provide opportunities to work with real problems, cooperatively with schoolmates and teachers, where one learns by one's own efforts, together with others, that 1/3 + 1/4 equals 7/12. In these contexts, many students develop some self-confidence and skill in thinking about numbers. We described several such programs above in which children do not hate mathematics because they know that mathematics is just thinking carefully about questions involving quantity. It also contributes to students approaching wide-ranging problems with reasoning and persistence. A constructive, experiential approach to learning arithmetic and algebra can be combined with practice, drill, and mastery of algorithms. But the reform curricula are not perfect. They are an important first step.

The greatest problem for math education in the United States is that there are nowhere nearly enough qualified math teachers. Moreover, those who are qualified don't usually work in the less affluent school districts. In order to provide quality math education for all public school students, teachers with math qualifications must be paid salaries comparable to those in business and industry. And learning math need not be limited to schoolrooms; afterschool programs and community organizing efforts can develop a new consciousness and pride, as illustrated by the Algebra Project.

But isn't there a contradiction between two different attitudes here? The Algebra Project of Bob Moses, which we admire and support, elevates the mastery of basic algebra to the level of a fundamental civil right, an actual demand on behalf of all children, especially the dispossessed inner-city children now so badly served in U.S. public schools. On the other hand, here we are arguing that algebra is not important or necessary for all people. Which is right?

It is unrealistic and unnecessary to guarantee that every child pass 10th grade algebra, and have a good facility with quadratic equations or with systems of two and three linear equations. What is necessary and realistic is that every child have the *opportunity* to learn algebra, in a well-equipped classroom, from a qualified, highly motivated teacher. All learners need to be able to think critically, with enough self-confidence to venture into the mathematical topics that are alienating when taught mechanically. Math phobia, which is frequently the result of the humiliating treatment of students, limits their ability to manage their financial lives and makes them vulnerable to deceptive lending practices.

In addition, it is essential to explicitly acknowledge the big difference between the school system in the United States and in many other countries. Here in the United States, vocational or industrial education in the public schools has almost been discredited. It was perceived as intrinsically discriminatory, a dumping ground for children in disadvantaged ethnic groups, especially Blacks and Latinos. There is an emerging goal that every child should be encouraged to have a college education. This goal is made hard to achieve by the discriminatory treatment at the primary and secondary levels. Such discrimination comes about largely as a consequence of the system of financing public education in the United States. Funding comes mainly from locally imposed property taxes. Educational districts in depopulated rural areas or in economically depressed urban areas have a much smaller tax base than those in affluent urban or suburban neighborhoods. As a consequence, less money is available to maintain the schools, and less money is available to attract the best teachers. So there is a huge disparity in the quality of public education between the inner city, for example, and the exclusive expensive suburbs of New York or Washington. A massive change in this system is urgently needed, and there is hope that it can be achieved under the new president, despite the present conditions of economic depression.

As we confront the many conflicting pressures that impact the U.S. educational system, we need to go beyond the polarity of reform and antireform issues. We need to find ways to make mathematics accessible and interesting to the greatest number of students while refraining from using it as a means of separating promising learners from alienated ones. By making mathematics more accessible and relevant to daily life and by having teachers who are passionate about the subject, we may succeed in decreasing the number of individuals who see themselves as incapable of dealing with numbers and numerical abstractions and help all learners to practice careful reasoning.

Math in College

At the university level, we believe that no one can be considered educated who doesn't have some appreciation of mathematical thinking and its importance to science. However, the math requirements in college usually serve no such purpose. Most students are put into what is called a precalculus sequence—namely, a review of high school algebra and trigonometry—even though most of them stop there and never take calculus. Of those who do take calculus, most do so only because it is required for admission to medical school or business school. These prerequisites for schools of law, medicine, or architecture should be reconsidered. When we ask the faculty of those schools, "What do you want your students to know about calculus?" we are repeatedly told, "It doesn't matter!" This "math filter" may have the advantage of being objective, easier to defend against charges of bias or favoritism. Still, it would be more rational and equitable to assess applicants to medical or business school according to abilities and commitments that are actually appropriate to their potential profession. For example, most doctors need to be able to make diagnoses. For that purpose, they need to elicit information from patients, to absorb and apply the results of medical research, and to develop and monitor adequate treatment. Thus, it seems that verbal and interpersonal intelligences and interest in science are more important for success in medicine than a knowledge of calculus. To select medical students we should look at portfolios, early internships, content-specific reasoning tasks, and judgments of their motivation for their chosen profession. The calculus filter is counterproductive. The kind of mathematics that physicians or entrepreneurs really are likely to use—basic statistics, and calculating with ratios and proportions—could be provided in a course specifically designed for them, either at the undergraduate or the professional level.

The following anecdote may be apocryphal, but it is enlightening:

An American mathematician of some note was returning from a trip abroad and had to go through customs. The U.S. customs officer asked him what he had been doing during his one week sojourn. The reply was that he had been at a mathematics conference. The customs officer then took this man aside and detained him for some time with a great many tedious questions about exactly where he had been and what he had been doing during his travels. The mathematician kept glancing nervously at his watch, worried that he would miss his connecting flight. The customs officer finally got to a point of asking our friend what he had for dinner each day. Finally the mathematician threw up his hands and exclaimed, "Why are you doing this to me?" The customs officer smiled and said, "Ah. Now you know how I felt when I took calculus."[41]

Maybe the professor finally knew how the trapped student felt. But did the former student know how the professor felt? Does any mathematician enjoy forcing his subject down the throats of passive victims who want only a passing grade and to escape from the endless lectures? Paul Halmos wrote:

> A class of students who take a course because of such requirements is a sad and discouraging class . . . the first prerequisite for the learning process to be both pleasant and effective—namely curiosity—is lacking, and that ruins it all. It ruins the teaching, it ruins the learning, it ruins the fun. I dream of the ideal university, full of students who are full of intellectual curiosity. The subset of those among them who take the mathematics course do so because they want to know mathematics . . . they come to me free, willing, and ask me to teach it to them. Oh, joy! If that really happened, I'd jump at the chance.[42]

Does it have to be this way? Could it be different? Nel Noddings dares to have such a dream. After 23 years as an elementary and high school teacher and an administrator in New Jersey public schools, Noddings earned a Ph.D. and became professor of education at Stanford from 1977 to 1998. She also served as acting dean of the School of Education. She is a former president of both the Philosophy of Education Society and the John Dewey Society. She has raised 10 children. She presents an alternative approach that recognizes diversity in interests, talents, plans, and hopes among learners.

She writes, "We are overly reluctant to face the fact that human interests vary widely and that many highly intelligent people are just not attracted to mathematics. . . . I don't know what talents and interests are lost under such coercion, what levels of confidence are eroded, what nervous habits develop, what

Figure 9-2. Nel Noddings, American educational philosopher. Courtesy of Nel Noddings.

rationalizations are concocted, or what evils are visited on the next generation as a result of our benevolent insistence."[43] We agree with her, that "tracks" should not be lower and higher, just different and that all honest work should receive respect and dignity.

Nor should we push all students to "think like a mathematician." Better let them learn, as she says, to use math for their own purposes.[44] We should reject any assumptions about shared universal capabilities and recognize the diversity of intellectual strength. Noddings even wrote, "I do not think that children who are poor at math, who may never—no matter how hard they try—understand algebra and geometry, are in any impor-

tant sense, handicapped, inferior, or in need of heroic intervention."[45] She asks, "Why should a student who wants to major in
literature, art, drama, law enforcement, history, or social work
'learn' algebra and geometry? . . . I have come to suspect that
teaching everyone algebra and geometry is both wasteful and
inconsiderate."

We could then be more attentive and helpful to students gen-
uinely interested in math. With them, she writes, we could even
"work for deeper self-understanding, discuss the loneliness that
sometimes accompanies extended intellectual work and the joy
that emerges from successful encounters with mathematics."[46]
We agree with her, that such students should understand that
their gifts are not higher than others—just different.

What Noddings is asking for on behalf of the student is
exactly what Halmos is asking for on behalf of the teacher!
She asks us to let students follow their interests and intellectual strengths, in addition to acquiring the mathematical skills
needed for daily existence and effective reasoning. We agree
with her position. Letting students deepen their knowledge in
diverse contexts, using their different interests, is basic to an
education that "counts."

Today, algebra in high school and calculus in college are the
primary "filters" for entering higher education. To justify this,
it is argued that citizens must be able to reason logically, as
voters and as consumers. Is it really true that passing algebra
or trigonometry proves that the student can reason logically
about politics or purchasing? We know of no evidence that
high school algebra or trigonometry increase the ability to criti-
cize phony advertising or political slogans. Some have a better
chance to learn critical thinking in a domain where they have an
interest, be it empirical science, literary analysis, or law. In non-
school settings logic is needed for carefully researching major
purchases, predicting the winners of sports events, or making
important life decisions.

We are relying on the view of intelligence advocated by Howard Gardner, which recognizes the variety of human cognitive

strengths and weaknesses. Gardner's theory of multiple intelligences describes the human mind as "a series of relatively separate faculties, with only loose and non-predictable relations with one another."[47] This opposes the traditional idea of intelligence as a unitary quality measured by a biologically predetermined IQ score. Gardner's theory draws evidence from two research areas. One is stroke victims, who lose some cognitive skills but retain others. The second area is patients with Williams' syndrome, children who excel in performing music but lack any ability to recognize other people's emotions. In his book *Frames of Mind*, Gardner includes on his list of intelligences: linguistic intelligence, which is revealed in expressive written and oral speech; logical mathematical intelligence, shown in detecting patterns, thinking logically, and carrying out mathematical operations; spatial intelligence, in recognizing and manipulating patterns both in wide open and confined spaces; musical intelligence, in identifying pitches, tones, and rhythms and using them for performance or composition; and bodily kinesthetic intelligence, which is possessed by athletes, dancers, and surgeons. In addition to all these, he specifies two emotional intelligences: "interpersonal" intelligence, shown by someone such as a teacher or counselor who works well with others; and "intrapersonal" intelligence, shown by someone whose self-knowledge guides his or her own life.

Most activities combine two or more modes of intelligence. In addition to bodily kinesthetic capability, a surgeon needs reasoning skills and a strong ability for visual representation. Mathematicians differ in how much they rely on logic and linguistic processes versus kinesthetic and visual ones. Most mathematics teaching neglects this diversity and thereby adds to the fear of failure felt by so many learners. At present, students are called "good at math" if they are good in the narrowly restricted ways it is presently taught. "Maybe if we really understood how different people use different cognitive capacities to solve problems we could design instruction so that many people were good at math."[48]

Gardner's different intelligences might be called abilities. Effective schooling needs to link students' learning to their diverse predispositions, interests, and cultures. If mathematics beyond the elementary level were studied only by students who are interested and motivated—perhaps only one out of every four—there would still be enough graduates for the jobs that need such knowledge. Those motivated to study mathematics in depth could then receive more sustained and effective instruction.

The reform efforts now going on in elementary schools do lessen math phobia. Such programs as the Algebra Project increase access to advanced courses and professional careers for minorities. These programs view learners as active participants in their education. They take time to achieve results; they require material resources, and a great investment of time in teacher training and individual tutoring.

Learning requires passion, joy, surprise, sustained interest, and the ability to get help from teachers and mentors. A student whose skill in mathematics is limited in early years may later be better able to persevere after he gains self-confidence in domains better matched to his ability.

The current dominant model of math instruction is mechanical and inflexible and results for many in lifelong avoidance of mathematics. More positive attitudes, and slow, cumulative gains in mathematical achievement, could come from *long-term* applications of innovative curricula. Instead of stark choices between traditional and reform approaches, we support the sustained improvement of mathematics teaching. The greatest likelihood of success is where local leadership is supportive and university assistance is available. University faculty and students in mathematics departments and math education programs can make important contributions to local public education.

Improving the teaching of mathematics is not limited to curriculum. Even in the early grades and junior high school, it can benefit from communitywide participation. Teaching by older students, as well as adult tutors, can be crucial in giving children the personal attention they need when they struggle with the abstract concepts of mathematics. Sharing everyday uses of arithmetic with parents and community members makes these concepts more accessible to young learners. If we focus on slow, cumulative gains, rather than on international competition and punitive, standardized testing, we can create more confidence and comfort at the early levels and provide more choices for students once they enter high school. For those who avoid mathematics classes in high school and college that don't seem relevant to their interests, we should provide opportunities to acquire relevant mathematics at a later stage. More mature individuals are more ready to take risks and to see the pragmatic value of math in the growing fields of technology.

Conclusions

We have presented many reasons for the widespread avoidance of mathematics by school children and by adults. Foremost among these is the formulaic way in which most teachers present mathematical abstractions. Many students become insecure and may avoid math for the rest of their lives. While some of these problems are being addressed by reform efforts in the United States and elsewhere, the pace of reform is slow and limited to curriculum. In our view, broader changes are needed to address this problem. Among these is discontinuing the use of mathematics as an academic filter.

Instead, the goal is to treasure diversity in talent and interest; to provide advanced mathematics teaching/learning to motivated students, while decreasing the number who suffer from math phobia. The challenge is to develop a systematic, society-wide perspective, rather than imposing the same values and approaches on both enthusiastic and reluctant learners. Because we love mathematics, we want to minimize the number of those who hate it. Our purpose in these proposals is to shift the premise of the current debate. It is to create a humanistic role for mathematics and its teaching in our culture, a way of teaching mathematics that focuses on the needs and abilities of students as well as society.

Bibliography

Carraher, T. N., Carraher, D., & Schliemann, A. D. (1985). Mathematics in the streets and in the schools. *British Journal of Developmental Psychology 3*, 21–29.

Charbonneau, M., & John-Steiner, V. (1988). Patterns of experience and the language of mathematics. In R. Cocking & J. P. Mestre (Eds.). *Linguistic and cultural influences on learning mathematics*. Hillsdale, N.J.: Erlbaum, pp. 91–100.

Cohen, R. (2006). What is the value of algebra? *Washington Post*. February 16, 2006.

Confrey, J. (1995). Student voice in examining "splitting" as an approach to ratio, proportions and fractions. Proceedings of PME 19, Recife, Brasil.

Cornell, C. (1999). I hate math! I couldn't learn it, and I can't teach it! *Childhood Education 75*(4), 225–230.

Davis, P. J. (2006). Mathematics and common sense. A case of creative tension. Wellesley, Mass.: A. K. Peters.

Devlin, Keith. (2009). MAA On-Line, January 2009. Should children learn math by starting with counting? Devlin's angle, http://www.maa.org/devlin/devlin_01_09.html

Dudley, U. (1997). Is mathematics necessary? *College Mathematics Journal 28*(5), 361–365.

Fosnot, C. T., & Dolk, M. (2001). *Young mathematicians at work*. Portsmouth, N.H.: Heinemann.

Gardner, H. (1993). *Frames of mind: The theory of multiple intelligences*. New York: Basic Books.

Gardner, H. (1999). *Intelligence reframed: Multiple intelligences for the 21st century*. New York: Basic Books.

Gerdes, P. (2001). On culture, geometrical thinking and mathematics education. In A. B. Powell & M. Frankenstein (Eds.). *Ethnomathematics: Challenging Eurocentrism in mathematics education.* Albany, N.Y.: State University of New York Press, pp. 223–246.

Gilman, L. (2001). The theory of multiple intelligences: Human intelligence. The Theory of Multiple Intelligences web Site. http://indiana.edu/~intell/mitheory.shtml.

Halmos, P. (1985). *I want to be a mathematician.* Washington, D.C.: Mathematical Association of America.

Hardy, G. H. (1948). *A mathematician's apology.* New York: Cambridge University Press.

Krantz, S. (2002). *Mathematical apocrypha.* Washington, D.C.: Mathematical Association of America.

Lave, J. (1988). *Cognition in practice: Mind, mathematics and culture in everyday life.* New York: Cambridge University Press.

Lester, W. (2005). Hate mathematics? You are not alone. Associated Press, August 16, 2005.

Levin, T. (2006). As math scores lag, a new push for the basics. *New York Times*, November 14 2006.

McCarthy, C. (1991). Who needs algebra? *Washington Post*, April 20, 1991.

Moses, R. P., & Cobb, C. E. Jr. (2001). *Radical equations: Math literacy and civil rights.* Boston: Beacon Press.

Moses, R., West, M. M., & Davis, F. E. (2009). Culturally responsive mathematics in the Algebra Project. In B. Greer, S. Thukhopadhyay, A. B. Powell, & S. Nelson-Barber. *Culturally responsive mathematics education.* New York: Routledge.

Noddings, N. (1993). Excellence as a guide to educational conversation. *Teachers College Record 94*(4), 730–743.

Noddings, N. (1994). Does everybody count? *Journal of Mathematical Behavior 13*(1), 89–104.

Noddings, N. (2003). *Happiness and education.* New York: Cambridge University Press.

Noddings, N. (2007). *The challenge to care in schools.* New York: Teachers College Press, pp. 151–159.

Pearson, R. S. (1991). Why don't most engineers use undergraduate mathematics in their professional work? *UME Trends 3:3* p. 8.

Piaget, J. (1965/1941). *The child's concept of number*. New York: W.W. Norton.

Rota, G. C. (1997). *Indiscrete thoughts*. Boston: Birkhäuser.

Russell, B. (1957). "The study of mathematics," in *Mysticism and logic*. New York: Doubleday.

Schliemann, A. D., Carraher, D.W., & Ceci, S. J. (1997). Everyday cognition. In J.W. Berry, P. R. Dasen, & T. S. Sarawathi (Eds.). *Handbook of cross-cultural psychology* (2nd ed.), vol. 2: Basic processes and developmental psychology. Boston: Allyn & Bacon, pp. 177–215.

Schmittau, J. (2003). Cultural historical theory in mathematics education. In A. Kozulin, B. Gindis, V. S. Ageyev, & S. N. Miller (Eds.). *Vygotsky's educational theory in cultural context*. New York: Cambridge University Press, pp. 225–245.

Schwartz, L. (2001). *A mathematician grappling with his century*. Boston: Birkhauser.

Slocum, J., & Sonneveld, D. (2006). *The 15 puzzle*. Beverly Hills, Calif.: Slocum Puzzle Foundation.

Tobias, S. (1993). *Overcoming math anxiety*. New York: W. W. Norton.

Treisman, U. (1991). Studying students studying calculus: A look at the lives of student mathematicians. *College Mathematics Journal 23*, 362–372.

Umland, K. (2006). Personal communication. Department of Mathematics and Statistics, University of New Mexico.

Umland, K., and Hersh, R. (2007). Mathematical discourse: The link from premathematical to fully mathematical thinking. *Philosophy and Education 19*, 1–10.

Walkerdine, V. (1997). Difference, cognition, and mathematics education. In A. B. Powell & M. Frankenstein (Eds.). *Ethnomathematics: Challenging Eurocentrism in mathematics education*. Albany, N.Y.: State University of New York Press, pp. 201–214.

Zaslavsky, C. (1996). *The multicultural math classroom*. Portsmouth, N.H.: Heinemann.

Conclusions

We have completed our journey, our tour around various aspects of mathematical life. We looked at the beginnings of mathematical life for children and students. Then we studied some of its special features as a unique subculture of modern society. We saw its ability, on the one hand, to provide its devotees with solace and refuge; and we saw, on the other hand, its dangers in permitting them isolation and eccentricity, which have in some rare cases extended to utter madness. We looked next at some of the glue that holds the mathematical community together, in a chapter on friendships, a chapter on communities, and a chapter on aging. Then our last two chapters turned to teaching, learning, and schooling.

Such a broad and inclusive look at mathematical life is hardly intended to prove a theory or preach a moral. But we can point to a few important issues. One is an obvious but often overlooked psychological fact: mathematical work, like every other kind of deeply engaged intellectual or artistic work, is deeply emotional. It relies on intense motivation; it brings with it elation and disappointment, happiness and grief.

Some feelings about clarity and certainty, and the pursuit of answers to long unresolved problems, are specific to mathematics. Other emotions are shared across disciplines. These include the pleasures of mentoring, the challenge of teaching, and the rewards of participating in a caring community, as well as the discomfort of competition for prizes and fame. The joy of discovery, an emotion that mathematicians share with scientists and artists, was thus celebrated by Paul Halmos: "The joy of suddenly learning a former secret and the joy of suddenly discovering a hitherto unknown truth are the same to me—both have the flash of enlightenment, the almost incredibly enhanced vision, and the ecstasy and euphoria of released tension."[1] But mathematicians are particularly vulnerable to a sense of inadequacy in a profession that remembers and honors so many of its most illustrious contributors.

The applied mathematician Fern Hunt has said, "No matter how good you actually are, there is definitely somebody who can run rings around you. . . . This and the fact that mathematics is a field a lot of people have trouble with causes a great deal of anxiety both within and outside the profession."[2]

We have met Lipman Bers before in this book, as a beloved mentor at the Courant Institute and as a leader among mathematicians struggling against abuses of human rights He expressed some of these emotions in an interview.[3]

Question: "When you say that mathematics is a very cruel profession, do you mean because the standards are so high?"

Bers: "The standards are high, and you never know whether you will be able to hack it. First you are afraid that you won't be able to understand your professors. Then you are afraid that you won't be able to write a thesis."

Bers was then asked whether, in spite of doubts, one does at last realize that one has succeeded. He answered: "If you have done something, yes. Nothing can compare with this pleasure! But then you start worrying—will you be able to do it again?"[4]

We have repeatedly mentioned the appeal of clarity and elegance that many future mathematicians find alluring. The usefulness of mathematics to physics, engineering, biology, and other disciplines also is a great motivation and satisfaction for future mathematicians.

But ambiguity, contradiction, and paradox are also inherent in mathematics. Life is ambiguous and contradictory. Mathematics is part of life. In so far as a philosophy of mathematics describes the total mathematical situation—process as well as content—naturally, it is also bound to be ambiguous. As William Byers writes, "Logic moves in one direction, the direction of clarity, coherence, and structure. Ambiguity moves in the other direction, that of fluidity, openness, and release. Mathematics moves back and forth between these two poles. . . . It is the interaction between these different aspects that gives mathematics this power."[5]

Mathematical culture includes not only known results and theorems but also open problems. These challenges call forth some of the feelings Bers identified—doubts and questions, as well as the pleasure of reaching a solution. Problems may be ambiguous. Working on them requires one to live with the tension of uncertainty. Mathematicians cherish stories about one of their heroes' arduous journey to resolve some long-standing conjecture. Such stories are told and retold as part of mathematical history and culture.

One powerful theme in this book has been the need for balance between single-minded absorption, on the one hand, and intellectual and emotional breadth, on the other. Part of the fascination of mathematics is its clarity, aesthetic appeal, and precision. But total immersion in these aspects can lead to a teaching style devoid of humor, lightness of touch, or compassion. It can even endanger someone who is vulnerable to obsession. For many mathematicians, a counterweight to immersion in their intellectual work is provided by love of friends and family or by joy in music or in nature. We think of the chamber music players of Göttingen and New York, and of the many hikers, swimmers, tennis players, collectors of butterflies or minerals, and lovers of music or poetry.

The second conclusion to this work is that despite its appearance of being individual and solitary, mathematical life is social as well as emotional. Every bit of mathematical work, whether problem solving, theory building, or practical application, takes its meaning and value from its interest and relevance to the mathematical community and to the larger society. This recognition goes counter to the stereotype of mathematics as an extreme academic ivory tower, as a sort of closed subculture disconnected from the concerns of socially oriented scholars or the public at large. By looking at controversies related to race, gender, age, and prize competitions, we have seen mathematical life entangled with the challenges and conflicts of contemporary culture. At the same time, immersion in mathematics has offered a temporary haven from war, persecution, and injustice.

Because it is social, there is always an ethical aspect to mathematical life. What you do affects others and can be helpful or harmful. In general, as part of a school or a university or even a corporation or a bureaucracy, a mathematician like anyone else may be competitive or cooperative, constructive or destructive, helpful or harmful. Moreover, since mathematics is integrally connected—financially, politically, and ideologically—to the larger society, the role a mathematician plays in his or her own professional community—for or against freedom of thought, social advancement, and human welfare—is subject to the same ethical judgments that would apply in any other realm of social life.

Applied mathematicians, who collaborate with physicists, biologists, or engineers, are judged by the usefulness of their work in real-world terms. Will they contribute to sophisticated methods of destruction? Or will their work be used for the benefit of humankind?

But the special activities of mathematicians are teaching and research. For a teacher, the main question is, What do you do for your students? Do you help to overcome the alienation so many young people feel when confronted with this rigorous discipline? Do you share with them your passion for the beauty of this discipline? Do you share your own turmoil when a solution is evasive?

As part of an ethical approach to the teaching of mathematics we have argued against using it as a filter—using it to decide who may get into graduate or professional programs. Mathematical knowledge is relevant for engineering. But for fields such as medicine, there are more appropriate ways to choose future doctors.

What are ethical concerns for mathematical researchers? Bill Thurston writes that the goal of mathematical research is *to advance mathematics*—not just to pile up theorems and proofs attached to your name. Do you try to make it possible for others as well as yourself to make big discoveries?

For those like G. H. Hardy who see themselves above all as artists, it is appropriate to be evaluated in the way one evaluates a composer or a novelist. Do you stay with the multitude? Or do you follow your own vision where it leads, even away from the most popular and acceptable trend? Do you go for the easy product for a guaranteed payoff without too much time and struggle? Or do you take on the most demanding project of which you are capable?

Mathematics is part of the broad tapestry of human thought. Like other parts of art and science, it is a search for pattern, harmony, proportion, and application. It offers dangers and frustrations, unreasonable and impossible demands. It also offers intense and memorable pleasures and satisfactions. Frustrations and satisfactions, dangers and pleasures, all are part of a deeply demanding and rewarding way of life.

Bibliography

Albers, D. J., Alexanderson, G. L., & Reid, C. (Eds.) (1990). *More mathematical people: Profiles and interviews.* Boston: Birkhäuser.

Byers, W. (2007). *How mathematicians think: Using ambiguity, contradiction, and paradox to create.* Princeton, N.J.: Princeton University Press.

Halmos, P. R. (1985). *I want to be a mathematician.* New York: MAA Spectrum, Springer-Verlag.

Henrion, C. (1997). *Women in Mathematics: The addition of difference.* Bloomington, Ind.: Indiana University Press, p. 226.

Thurston, W. (2005). On proof and progress in mathematics. In R. Hersh (Ed.). *Unconventional essays on the nature of mathematics.* New York: Springer.

Review of the Literature

The "popular" literature on mathematics is booming. Every year excellent new books are added to the shelves. There are books that intend to teach a particular branch of mathematics. Textbooks of course are a whole industry unto themselves. There are books meant for the casual or "pleasure" reader, which deal with special topics beyond the standard high school or lower-division college level—lots of them on probability and statistics, and recently, good books about topology, about group theory, about graphs and networks, about non-Euclidean geometry.

Then there are books about the history of mathematics, sometimes in the form of collections of brief biographies arranged chronologically. There are books about the philosophy of mathematics. There are lots of books about teaching and education, even apart from those specifically meant for the professional training of teachers. In particular, books about geniuses and prodigies, and books about math anxiety and math avoidance and even "math for dummies." And finally there are biographies and autobiographies of mathematicians. It is from this last group that much of the material in this book has been taken. Of course, in writing this book we have tried to look at all possible sources. So here we would like to offer you our overview and evaluation of the many, many books we have used.

First of all, violating modesty, let us mention *The Mathematical Experience* by Philip J. Davis, Reuben Hersh, and Elena Marchisotto. This book is now over 20 years old but is still useful. Its main intention was to remove the curtain or veil that hid the life and thinking of mathematicians from the reading public. In doing so, it accomplished something novel: presenting in reasonably accessible form some of the latest theories and results from pure mathematics, including logic, harmonic analysis, and group theory, to the general public. It was a breakthrough that encouraged many other authors to undertake similar rash enterprises.

We list below a few recent books in several different categories that we found interesting and readable. But the main purpose is to list the biographies and autobiographies, with special notice for the ones we consider outstanding. (Not all of them have actually been referred to in this book.)

Biography Collections

The honors class: Hilbert's problems and their solvers, Benjamin H. Yandell. Natick, Mass.: A.K. Peters, 2002.
    This history of the research inspired by Hilbert's famous list of 23 problems is invaluable.
Men of mathematics, E.T. Bell. New York: Simon and Schuster, 1965.
    A standard classic, not a totally reliable history text but intensely readable and extremely widely read by past, present, and future mathematicians.
Remarkable mathematicians: From Euler to von Neumann, Ioan James. Washington, D.C.: Mathematical Association of America, 2002.
Driven to innovate: A century of Jewish mathematicians and physicists, Ioan James. Witney, Oxfordshire: Peter Lang, 2009.

Autobiography

The apprenticeship of a mathematician, André Weil. Translated from the French by Jennifer Gage. Boston: Birkhäuser, 1992.
    André Weil's autobiography goes only as far as his midcareer and has hardly any mathematics, but its flavor of casual presumption of unquestioned supremacy does communicate the flavor of this unique personality.

Ex-prodigy: My childhood and youth, Norbert Wiener. New York: Simon and Schuster, 1953.

I am a mathematician: The later life of a prodigy. Garden City, N.Y.: Doubleday, 1956. An autobiographical account of the mature years and career of Norbert Wiener and a continuation of the account of his childhood in *Ex-prodigy*.

A mathematician's apology, G. H. Hardy. With a foreword by C. P. Snow. New York: Cambridge University Press, 1992.

A mathematician grappling with his century, Laurent Schwartz. Translated from the French by Leila Schneps. Boston: Birkhäuser, 2001.

Enigmas of chance: An autobiography, Mark Kac. Berkeley, Calif.: University of California Press, 1987.

Adventures of a mathematician, S. M. Ulam. New York: Scribner, 1976.

The recollections of Eugene P. Wigner as told to Andrew Szanton. New York: Plenum Press, 1992.

The education of a mathematician, Philip J. Davis. Natick, Mass.: A. K. Peters, 2000.

I want to be a mathematician: An automathography, Paul R. Halmos. New York: Springer-Verlag, 1985.
Conveys well the unique flavor and personality of this famous American writer and teacher of mathematics.

Random curves, Neal Koblitz. New York: Springer, 2008.
Lively and controversial memoir of a leading algebraist, cryptographer, and activist educator.

Modern Biography

Perfect rigor, Masha Gessen. Boston: Houghton Mifflin Harcourt, 2009.
The life of Grisha Perelman, who proved Poincaré's conjecture.

Charles Sanders Peirce: A life, Joseph Brent. Bloomington, Ind.: Indiana University Press, 1993.

Courant in Göttingen and New York: The story of an improbable mathematician, Constance Reid. New York: Springer-Verlag, 1976.

Constance Reid's books set a high standard for literary quality and careful research. We have used her biographies of Hilbert, Courant, and her sister Julia Robinson.

Hilbert, Constance Reid. With an appreciation of Hilbert's mathematical work by Hermann Weyl. New York, Springer-Verlag, 1970.

Julia, a life in mathematics, Constance Reid. Washington, D.C.: Mathematical Association of America, 1996.

A convergence of lives: Sofia Kovalevskaya, scientist, writer, revolutionary, Ann Hibner Koblitz. Cambridge, Mass.: Birkhauser Boston, 1983

Alan Turing: The enigma, Andrew Hodges. Foreword by Douglas Hofstadter. New York: Walker, 2000.

Stephen Smale: The mathematician who broke the dimension barrier, Steve Batterson. Providence, R.I.: American Mathematical Society, 2000.

The man who knew infinity: A life of the genius Ramanujan, Robert Kanigel. New York: Scribner, 1991.

An outstanding literary achievement.

Ramanujan: Twelve lectures on subjects suggested by his life and work, G. H. Hardy. New York: Chelsea.

The wind and beyond: Theodore von Kármán, pioneer in aviation and pathfinder in space, Theodore von Kármán with Lee Edson. Boston: Little, Brown, 1967.

The man who loved only numbers: The story of Paul Erdős and the search for mathematical truth, Paul Hoffman. New York: Hyperion, 1998.

My brain is open: The mathematical journeys of Paul Erdős, Bruce Schechter. New York: Simon and Schuster, 1998.

Alfred Tarski: Life and logic, Anita B. Feferman and Solomon Feferman. New York: Cambridge University Press, 2004.

Politics, Logic, and Love: The life of Jean van Heijenoort, Anita Burdman Feferman. Wellesley, Mass.: A.K. Peters, 1993.

Van Heijenoort was murdered by an ex-wife, after a career as a historian of logic, following years of service as private secretary to Leon Trotsky.

Willard Gibbs, Muriel Rukeyser. New York: Dutton, 1964.

Logical Dilemmas: The life and work of Kurt Gödel, John Dawson. Wellesley, Mass.: A.K. Peters, 1997.

Incompleteness: The proof and paradox of Kurt Gödel, Rebecca Goldstein. New York: W.W. Norton, 2005.

Abraham Robinson: The creation of nonstandard analysis—A personal and mathematical odyssey, Joseph Warren Dauben. Princeton, N.J.: Princeton University Press, 1995.

Georg Cantor: His mathematics and philosophy of the infinite, Joseph Warren Dauben. Cambridge, Mass.: Harvard University Press, 1979.

The unreal life of Oscar Zariski, Carol Ann Parikh. Boston: Academic Press, 1991.
Readable and reliable.

King of infinite space: Donald Coxeter: The man who saved geometry, Siobhan Roberts. New York: Walker, 2006.

John von Neumann and Norbert Wiener: From mathematics to the technologies of life and death, Steve J. Heims. Cambridge, Mass.: MIT Press, 1980.

Henri Poincaré, critic of crisis: Reflections on his universe of discourse, Tobias Dantzig. New York, Scribner, 1954.

R. L. Moore: Mathematician and teacher, John Parker, Washington, D.C.: Mathematical Association of America, 2005.

Emmy Noether, 1882–1935, Auguste Dick. Translated by Heidi Blocher. Boston: Birkhäuser, 1981.

Count down: Six kids vie for glory at the world's toughest math competition, Steve Olson. Boston: Houghton Mifflin, 2004.

Logic's lost genius: The life of Gerhard Gentzen, Eckart Menzler-Trott. Providence, R.I.: American Mathematical Society, 2007.

Jean Cavaillès: A philosopher in time of war 1903–1944, Gabrielle Ferrieres. Studies in French Civilization, 16. Translated by T. N. F. Murtagh. Lewiston, N.Y.: Edwin Mellen Press, 2000.

Cavaillès was a logic professor at the Sorbonne who spent his afternoons organizing sabotage against the Nazi occupiers; fictionalized in the 1969 movie *Army of Shadows*.

Ernst Zermelo: An approach to his life and work, Heinz-Dieter Ebbinghaus and V. Peckhaus (contributor). New York: Springer, 2007.

Niels Henrik Abel and his times, Arild Stubhaug and Richard R. Daly. New York: Springer, 2000.

The Mathematician Sophus Lie, Arild Stubhaug and R. Daly. New York: Springer, 2002.

Classic Biography

Isaac Newton, James Gleick. New York: Pantheon Books, 2003.

The life of Isaac Newton, Richard S. Westfall. New York: Cambridge University Press, 1993.

Never at rest: A biography of Isaac Newton, Richard S. Westfall. New York: Cambridge University Press, 1980.

The mathematical career of Pierre de Fermat (1601–1665), Michael S. Mahoney. Princeton, N.J.: Princeton University Press, 1973.

Blaise Pascal: Mathematician, physicist, and thinker about God, Donald Adamson. New York: St. Martin's, 1995.

Joseph Fourier: The man and the physicist, John Herivel. Oxford: Clarendon Press, 1975.

Joseph Fourier, 1768–1830: A survey of his life and work, I. Grattan-Guinness in collaboration with J. R. Ravetz. Based on a critical edition of his monograph on the propagation of heat, presented to the Institut de France in 1807. Cambridge, Mass.: MIT Press, 1972.

Life of Sir William Rowan Hamilton, Robert Perceval Graves. New York: Arno Press, 1975.

Sir William Rowan Hamilton, Thomas L. Hankins. Baltimore: Johns Hopkins University Press, 1980.

Carl Friedrich Gauss: Titan of science, G. Waldo Dunnington. With additional material by Jeremy Gray and Fritz-Egbert Dohse. Washington, D.C.: Mathematical Association of America, 2004.

Carl Friedrich Gauss: 1777–1977, Karin Reich. Translated by Patricia Crampton Bonn-Bad Godesberg: Inter Nationes, 1977.

Euler: The master of us all, William Dunham. Washington, D.C.: Mathematical Association of America, 1999.

Sophie Germain: An essay in the history of the theory of elasticity, Louis L. Bucciarelli and Nancy Dworsky. Boston: D. Reidel, 1980. Sold and distributed in the United States and Canada by Kluwer, Boston.

History

A history of mathematics: An introduction, Victor J. Katz. Boston: Addison-Wesley, 2009.

Mathematicians under the Nazis, Sanford L. Segal. Princeton, N.J.: Princeton University Press, 2003.

Golden years of Moscow mathematics, Smilka Zdravkovska and Peter L. Duren (Eds.). Providence, R.I.: American Mathematical Society, 1993.

An eyewitness account of an inspiring and tragic episode.

A century of mathematics in America, Peter Duren, (Ed.). With the assistance of Richard A. Askey, Uta C. Merzbach. Providence, R.I.: American Mathematical Society, 1988–1989.

Mathematics at Berkeley: A history, Calvin C. Moore. Wellesley, Mass.: A.K. Peters, 2007.

Mathematical thought from ancient to modern times, Morris Kline. New York: Oxford University Press, 1990, 1972.

Origin of Chaos

Poincaré and the three body problem, June Barrow-Green. Providence, R.I.: American Mathematical Society, 1997.

The chaos avant-garde: Memories of the early days of chaos theory, Ralph Abraham and Yoshisuke Ueda (Eds.). River Edge, N.J.: World Scientific, 2000.

Chaos: Making a new science, James Gleick. New York: Viking, 1987.

MISCELLANEOUS

The mathematical experience, Philip J. Davis and Reuben Hersh. With an introduction by Gian-Carlo Rota. Cambridge, Mass.: Birkhäuser, 1980.

How mathematicians think. William Byers. Princeton, N.J.: Princeton University Press, 2007.

Mathematics: Frontiers and perspectives, V. Arnold et al. (Eds.). Providence, R.I.: American Mathematical Society, 2000.

Mathematical people: Profiles and interviews, Donald J. Albers and G. L. Alexanderson (Eds.). Introduction by Philip J. Davis. Boston: Birkhäuser, 1985.

More mathematical people: Contemporary conversations, Donald J. Albers, Gerald L. Alexanderson, and Constance Reid, (Eds.). Boston: Harcourt Brace Jovanovich, 1990.
  The two volumes, *Mathematical People* and *More Mathematical People,* have fascinating interviews with many living mathematicians.

The mathematician's brain, David Ruelle. Princeton, N.J.: Princeton University Press, 2007.

Littlewood's miscellany, Béla Bollobás (Ed.). New York: Cambridge University Press, 1986.

Radical equations: Math literacy and civil rights, Robert P. Moses and Charles E. Cobb, Jr. Boston: Beacon Press, 2001.
  Bob Moses, with the assistance of Charles Cobb, tells his story and presents his theories and goals.

Women in mathematics: The addition of difference, Claudia Henrion. Bloomington, Ind.: Indiana University Press, 1997.

Math education at its best: The Potsdam model, D. K. Datta. Framingham, Mass.: Center for Teaching/Learning of Mathematics, 1993.

Complexities: Women in mathematics, Bettye Anne Case and Anne M. Leggett (Eds.). Princeton, N.J.: Princeton University Press, 2005.

Indiscrete thoughts, Gian-Carlo Rota. Boston: Birkhäuser, 1997.

Jackson, A. (1999). The IHEs at forty, *Notices of the American Mathematical Society* 46(3), 330.

Récoltes et semailles, Alexandre Grothendieck. First 100 pages translated by Roy Lisker (2007). Conditions for obtaining these pages can be read at the Grothendieck circle web site http://www.grothendieckcircle.org July 20, 2007.

A document unique in the history of science.

Lisker, R. (1990). Ferment vol. V(5), June 25. The quest for Alexandre Grothendieck; #6 October 1: Grothendieck 2; #7 October 25: Grothendieck 3; #8 November 27; Grothendieck, 4: #9 January 1: Grothendieck 5. These are contained in a book entitled, *The Quest for Alexandre Grothendieck*, available from the author. Middletown, Conn.

You failed your math test, Comrade Einstein, Shifman, M. (Ed.). Singapore: World Scientific, 2005.

Mathematics under the microscope, Alexandre Borovik. Wordpress. http://micromath.wordpress.com. Also Providence, R.I.: American Mathematical Society, 2010.

The Number Sense, S. Dehaene. New York: Oxford University Press, 1997.

An introduction to understanding how mathematics sits in the human brain.

Biographies

NOTE: In this section we have sometimes gone into more detail about interesting people who are not so well known, and been satisfied with brief mention of very famous mathematicians. (This is not an exhoustive list of everyone mentioned in the book).

Much of this information was obtained by searching the World Wide Web (Internet). Thanks are due to Google and to the indispensable web sites Wikipedia, MacTutor (written and edited by J. J. O'Connor and E. F. Robertson at the University of St. Andrews), and Mathematicians of the African Diaspora (maintained by Scott W. Williams).

Ralph Abraham (1936–). U.S. mathematician actively involved in the development of the theory of dynamical systems in the 1960s and 1970s.

Jean d'Alembert (1717–1783). French mathematician who was a pioneer in the study of differential equations and their use in physics. He studied the equilibrium and motion of fluids.

P. S. Aleksandrov (1896–1982). Russian topologist who wrote about 300 scientific works in his long career. He laid the foundations of homology theory in a series of fundamental papers between 1925 and 1929.

Richard D. Anderson (1922–2008). U.S. mathematician who was a student of Robert Lee Moore. His work at first centered around the geometric topology of continua. He subsequently was largely responsible, along with his students, for developing infinite-dimensional topology.

Archimedes of Syracuse (c. 287 BC–c. 212 BC). Outstanding genius physicist and mathematician of the classical era.

Vladimir I. Arnold (1937–2010). Russian mathematician who while still a teen-aged student of Andrei Kolmogorov at Moscow State

University, solved Hilbert's 13th problem, by showing that any continuous function of several variables can be constructed with a finite number of two-variable functions. Since then he has made major contributions to an astounding number of different mathematical disciplines.

Michael Atiyah (1929–). Lebanese-British mathematician widely considered one of the greatest geometers of the 20th century. In the 1960s his path-breaking work with Isadore Singer produced the Atiyah-Singer index theorem, a result that helped to develop several branches of mathematics. Earlier, together with Friedrich Hirzebruch, he founded the study of another major tool in algebraic topology: topological K-theory. It was inspired by Alexander Grothendieck's work on the Riemann-Roch theorem and has since generated algebraic K-theory.

John Carlos Baez (1961–). U.S. mathematical physicist at the University of California, Riverside. He is known for his work on loop quantum gravity and on applications of higher categories to physics. His sister Joan is a famous singer.

Stefan Banach (1892–1945). Polish mathematician who founded modern functional analysis and made major contributions on topological vector spaces, measure theory, integration, and orthogonal series.

Henri Baruk (1897–1999). French neuropsychiatrist. Baruk spent his childhood living among patients at the asylum where his father Jacques was the director.

Edwin Beckenbach (1906–1982). U.S. mathematician who contributed to the creation of the Institute for Numerical Analysis at UCLA in 1948. Its SWAC computing machine was one of the half-dozen most powerful computers in the world. The mathematicians who gathered to use it made UCLA known around the world.

Edward Griffith Begle (1914–1978). U.S. mathematician who was the director of the School Mathematics Study Group (SMSG), the group mainly credited for developing the "new math."

Eric Temple Bell (1883–1960). Scottish-American number theorist and prolific author. His *Men of Mathematics* is a very widely read collection of mathematical biographies.

Felix Bernstein (1878–1956). German statistician and mathematician who taught at Göttingen from 1907 to 1934. In 1921 he founded the Institute of Mathematical Statistics, and in 1934 he emigrated to the United States. He returned to Göttingen in 1948. He published a famous theorem on the equivalence of sets while at Cantor's seminar at Halle in 1897.

Lipman Bers (1914–1993). Latvian-American mathematician who

worked on Riemann surfaces. He received his Ph.D. in 1938 from the University of Prague under Charles Loewner. He was a much loved and admired mentor of graduate students and an outstanding defender of human rights internationally.

Abram Samoilovitch Besicovitch (1891–1970). Russian-Jewish mathematician who studied under A. A. Markov at St. Petersburg University. He converted to Eastern Orthodoxy, joining the Russian Orthodox Church, on marrying in 1916. In 1924 he joined Harald Bohr in Copenhagen, where he worked on almost periodic functions, which now bear his name. He moved to Cambridge in 1927, where he was appointed to the Rouse Ball Chair of Mathematics, which he held until his retirement in 1958. He worked mainly on combinatorial methods and questions in real analysis, such as the Kakeya needle problem and the Hausdorff-Besicovitch dimension.

Enrico Betti (1823–1892). Italian mathematician who taught at the University of Pisa and was noted for contributions to algebra and topology. Betti also did important work in theoretical physics, in particular on potential theory and elasticity.

R H Bing (1914–1986). U.S. mathematician, student of Robert Lee Moore, who worked on general topology, particularly on metrization, and on planar sets, where he examined webs, cuttings, and planar embeddings.

George David Birkhoff (1884–1944). First leading U.S. mathematician educated in the United States, at Chicago and Harvard. His most important work was the ergodic theorem he proved in 1931.

Joan S. Lyttle Birman (1927–). U.S. mathematician. After years as a systems analyst in the aircraft industry, she took a break to raise three children. In 1961 she began working with Wilhelm Magnus and in 1968 received a Ph.D. from the Courant Institute of Mathematical Sciences. Birman's mathematical work has focused on low-dimensional topology: braids, knots, surface mappings, and 3-manifolds.

David Blackwell (1919–2010). Professor emeritus of statistics at the University of California, Berkeley, and one of the eponyms of the Rao-Blackwell theorem. In 1965 he was the first African American to be inducted into the National Academy of Sciences. David Blackwell said that the work that gave him the most satisfaction was infinite games and analytic sets. He found a game theory proof of the Kuratowski reduction theorem connecting the areas of game theory and topology.

André Bloch (1893–1948). French mathematician who is remembered for a result about univalent functions called Bloch's theorem. All his mathematical output was produced while he was confined in an institution for the criminally insane.

Lenore Blum (1942–). U.S. mathematician and logician teaching at Carnegie-Mellon. She described her work as follows: "Continuity is the mathematics of calculus and physics but there's never been a theory of computation that deals with this continuum."

Ralph Philip Boas, Jr. (1912–1992). U.S. mathematician, teacher, and journal editor at Northwestern University who wrote over 200 papers mainly in the fields of real and complex analysis.

Harald Bohr (1887–1951). Danish mathematician who worked on Dirichlet series and applied analysis to the theory of numbers. He is the only mathematician to win an Olympic medal (on Denmark's soccer team in 1908). He was the brother of the great physicist Niels Bohr.

Béla Bollobás (1943–). Hungarian-British graph theorist at Cambridge University. After earning a doctorate in discrete geometry in 1967, he spent a year in Moscow with I. M. Gelfand, and then in 1972 received a second doctorate, in functional analysis, from Cambridge.

János Bolyai (1802–1860). Hungarian mathematician who was one of the three famous discoverers of non-Euclidean geometry along with Gauss and Lobatchevsky.

Nicolas Bourbaki. Pseudonym of a group of (mainly) French mathematicians, established in 1935, who dominated much of pure mathematics in the 1950s and 1960s.

L. E. J. Brouwer (1881–1966). Dutch mathematician best known for his topological fixed point theorem. He founded the doctrine of mathematical intuitionism, which views mathematics as the formulation of mental constructions that are governed by self-evident laws.

Felix E. Browder (1927–). U.S. mathematician known for his work on elliptic partial differential equations. He is the brother of the mathematicians William and Andrew Browder. He received a doctorate from Princeton University in 1948. Browder was the recipient of the 1999 National Medal of Science. He also served as president of the American Mathematical Society from 1999 to 2000.

Justine Bumby. U.S. mathematician who lived with Alexandre Grothendieck. Their son is the statistician John Grothendieck.

Georg Cantor (1845–1918). German mathematician who founded set theory, which is considered by some to be the foundation of mathematics. He introduced the concept of infinite cardinal and ordinal numbers.

Lazare Nicolas Marguérite Carnot (1753–1823). French mathematician best known as a geometer. In 1803 he published *Géométrie de position* in which directed magnitudes were first systematically used in geometry.

Henri Cartan (1904–2008). French mathematician who worked on analytic functions, the theory of sheaves, homological theory, algebraic topology, and potential theory, producing significant developments in all these areas.

Pierre Cartier (1932–). French mathematician and member of Bourbaki. He stayed in the group until he retired in 1983. "I estimate that I contributed about 200 pages a year during all this time with Bourbaki."

Mary Cartwright (1900–1998). British specialist in differential equations. She was the first woman to receive the Sylvester Medal and to serve on the Council of the Royal Society. She was president of the London Mathematical Society in 1961–1962, so far the only woman president. Cartwright had a gift for going to the heart of a matter and for seeing the important point, both in mathematics and in human affairs.

Augustin-Louis Cauchy (1789–1857). French mathematician who pioneered the study of analysis, both real and complex, and the theory of permutation groups. He also did research in convergence and divergence of infinite series, differential equations, determinants, probability, and mathematical physics.

Arthur Cayley (1821–1895). English mathematician at Cambridge University. His most important work was on the algebra of matrices and on non-Euclidean and n-dimensional geometry.

N. G. Chebotaryev (1894–1947). Russian mathematician who taught in Odessa and Kazan. His "density theorem" was used in Emil Artin's solution of Hilbert's 9th problem (the most general law of reciprocity.)

Pafnuty Lvovich Chebyshev (1821–1894). Russian mathematician known for his work in probability, statistics, and number theory. He is considered a founding father of Russian mathematics. Among his students were the prolific mathematicians Dmitry Grave, Aleksandr Korkin, Aleksandr Lyapunov, and Andrei Markov. According to the Mathematics Genealogy Project, Chebyshev has about 5000 mathematical "descendants."

Shiing-Shen Chern (1911–2004). Chinese-American mathematician, who was one of the leaders in differential geometry in the 20th century.

Claude Chevalley (1909–1984). French mathematician who was a founding member of Bourbaki and a major influence on ring theory and group theory

William Schieffelin Claytor (1908–1967). U.S. mathematician. He worked as a researcher while teaching 18 to 21 hours per week and serving as chair of the department of mathematics at Howard University. In 1980 the National Association of Mathematicians instituted the Claytor Lecture Series in his honor.

Paul Cohen (1934–2007). U.S. mathematician who invented a new technique in set theory he called "forcing" and used it to prove the independence of the axiom of choice, and of the generalized continuum hypothesis. He spent most of his professional life at Stanford University.

Zerah Colburn (1804–1839). Famous U.S. child prodigy of the 19th century. Born in Cabot, Vermont, and educated at Westminster School in London, he was thought to be mentally retarded until the age of 7. When he was 7 years old he took 6 seconds to give the number of hours in 38 years, 2 months, and 7 days.

Jere Confrey. U.S. professor of mathematics education at Washington University. She was a cofounder of the UTEACH program for secondary math and science teacher preparation at the University of Texas in Austin and was the founder of the summer math program for young women at Mount Holyoke College and cofounder of Summer Math for Teachers.

John Horton Conway (1937–). English mathematician now at Princeton in the United States. He made major contributions to group theory, created a theory of "surreal numbers," and has done leading research in knot theory, number theory, game theory, quadratic forms, coding theory, and tilings.

Richard Courant (1888–1972). German-American who was the leader of the Mathematics Institute at Göttingen. After that institute was virtually destroyed by Hitler, he went to New York and created at NYU an Institute of Mathematical Sciences which became a world leader in research. In 1964 it was named the Courant Institute after him.

Harold Scott MacDonald Coxeter (1907–2003). Canadian mathematician who graduated from Cambridge and worked most of his life in Canada. He worked mainly in geometry, making major contributions to the theory of polytopes, non-Euclidean geometry, group theory, and combinatorics.

Crafoord Prize. Science prize established in 1980 by Holger Crafoord, a Swedish industrialist, and his wife, Anna-Greta Crafoord. Administered by the Royal Swedish Academy of Sciences, the prize "is intended to promote international basic research in the disciplines of Astronomy and Mathematics; Geosciences; Biosciences, with particular emphasis on ecology and Polyarthritis (rheumatoid arthritis)," the disease from which Holger Crafoord severely suffered in his last years. The prize is presented by the King of Sweden (who also presents the awards at the December Nobel prize award ceremony). The prize sum of US $500,000 (2007) is intended to fund further research by the prize winner.

Mihaly Csikszentmihalyi (1934–). Hungarian-American psychology professor at Claremont Graduate University in Claremont, California. He is noted for his work on the study of creativity and subjective well-being and is best known as the architect of the notion of "flow" and for his years of research and writing on the topic.

George Dantzig (1914–2005). U.S. mathematics professor whose seminal work on the simplex method of linear programming is the foundation for much of systems engineering and is widely used in network design and component design in computer, mechanical, and electrical engineering.

Tobias Dantzig (1884–1956). Baltic-German-Russian-American mathematician who was the father of George Dantzig and the author of *Number: The Language of Science*, an outstanding book presenting deep mathematics in a way accessible to the layperson.

Joseph Dauben. U.S. historian of mathematics. He is the author of books about Georg Cantor and Abraham Robinson.

Harold Davenport (1907–1969). English mathematician known for extensive work on number theory. From about 1950 on he led a group that was the successor to the school of mathematical analysis of G. H. Hardy and J. E. Littlewood but more devoted to analytic number theory. This implied problem solving and hard analysis. The outstanding work in diophantine approximation of Klaus Roth and Alan Baker showed what this can do. This emphasis on concrete problems contrasted sharply with the abstraction of Bourbaki, which was then active just across the English Channel.

Chandler Davis (1926–). American-Canadian mathematician whose research has been principally on linear algebra and operator theory in Hilbert space. He began his writing career with *Astounding Science Fiction* in 1946. He has been a long-standing faculty member at the University of Toronto and is editor of *The Mathematical Intelligencer*.

Karel de Leeuw (1930–1978). U.S. mathematician at Stanford University who specialized in harmonic analysis and functional analysis.

He received a doctorate from Princeton in 1954 under Emil Artin. He was murdered by Theodore Streleski, a Stanford doctoral student for 19 years whom he briefly advised.

Pierre Deligne (1944–). Belgian mathematician who was a student of Grothendieck who turned Grothendieck's philosophy of motives into the driving force behind many areas of current algebraic geometry and arithmetic. Deligne has brought about a new understanding of the cohomology of varieties.

Jean Delsarte (1903–1968). French mathematician and member of Bourbaki who was best known for his work on mean periodic functions and translation operators. He was an analyst of great power and originality.

Abraham de Moivre (1667–1754). A French mathematician, famous for de Moivre's formula, which links complex numbers and trigonometry, and for work on the normal distribution and probability theory. His parents were Protestants. Freedom of worship had been guaranteed in France since 1598 by the Edict of Nantes, but in 1685 Louis XIV revoked the Edict, leading to the expulsion of the Huguenots. De Moivre was imprisoned but eventually was able to emigrate to England. As a foreigner in England, he was never able to gain a university position, but lived in poverty on his earnings as a tutor. Nevertheless, he became a friend of Isaac Newton, Edmund Halley, and James Stirling, and was elected a Fellow of the Royal Society. Newton used to fetch him every evening for philosophical discourse from the coffee house where he could usually be found.

Augustus De Morgan (1806–1871). English mathematician and logician who defined and introduced the term "mathematical induction," thereby putting that method on a rigorous basis.

René Descartes (1596–1650). French philosopher whose work is often considered the origin of modern philosophy. In his book *La géométrie* he introduced the systematic application of algebra to geometry, the first major advance in geometry since ancient times.

Jean Dieudonné (1906–1992). French mathematician who was a founding member of Bourbaki. He drafted much of the Bourbaki series of texts and also became provost of the University of Nice.

James A. Donaldson (1941–). U.S. mathematician and dean at Howard University who has worked on differential equations and applied mathematics.

Joseph Doob (1910–2004). U.S. probabilist who won the Steele Prize for his fundamental work on establishing probability as a branch of mathematics and for his continuing profound influence on its development.

Johann Peter Gustav Lejeune Dirichlet (1805–1859). German mathematician who proved in 1837 that, in any arithmetic progression whose first term has no common factor with the difference, there are infinitely many primes.

Underwood Dudley (1937–). U.S. mathematics professor at De Pauw University in the United States.

Freeman Dyson (1923–). English-American physicist and mathematician famous for his collaboration with Richard Feynman on Feynman's method of path integrals. He is the author of many books on science and social problems.

Nikolai Vladimirovich Efimov (1910–1982). Russian geometer and administrator at Moscow State University.

Samuel Eilenberg (1913–1998). Polish-American mathematician who coauthored with Norman Steenrod the famous text *Foundations of Algebraic Topology* in 1952. At that time there were many different and confusing versions of homology theory. Eilenberg and Steenrod showed how to state all these different theories as functors of homology, mapping the category of pairs of spaces to the category of groups or rings and employing suitable axioms such as "excision." He had a second life as a collector and connoisseur of modern art.

Albert Einstein (1879–1955). German physicist who contributed more than any other scientist to the modern vision of physical reality. His special and general theory of relativity is the most satisfactory model of the large-scale universe.

Paul Erdős (1913–1996). Hungarian mathematician who was one of the leading advocates and creators of discrete mathematics. He was peerless at both posing and solving problems in combinatorics, number theory, and other areas.

Alexander Sergeyevich Esenin-Volpin (1924–). Russian-American poet and mathematician. In the former Soviet Union, he was a leader of the human rights movement and spent 14 years incarcerated in prisons, in *psikhushkas*, and in exile.

Euclid of Alexandria (c. 325 BC–c. 265 BC). Euclid's treatise on geometry, *The Elements*, dominated Western mathematics education for more than 2000 years.

Leonhard Euler (1707–1783). Swiss mathematician who made enormous contributions to a wide range of mathematics and physics including analytic geometry, trigonometry, geometry, calculus, number theory, fluid dynamics, and elasticity.

Daniel Leonard Everett (1951–). U.S. linguistics professor best known for his study of the Amazon Basin's Pirahã people and their language.

Gerd Faltings (1954–). German mathematician known for his work on arithmetic algebraic geometry. He was awarded the Fields Medal in 1986 for proving the Mordell conjecture, which states that any nonsingular projective curve of genus $g > 1$ defined over a number field K contains only finitely many K-rational points.

Pierre de Fermat (1601–1665). French lawyer and government official remembered for his work on number theory, in particular on Fermat's Last Theorem, and for important contributions to the foundations of the calculus and probability.

Fields Medal. Award given to two, three, or four mathematicians not over 40 years of age at each International Congress of the International Mathematical Union, a meeting that takes place every 4 years. The Fields Medal is widely viewed as the top honor a mathematician can receive. It comes with a monetary award, which in 2006 was $15,000 (Canadian). It was founded at the behest of Canadian mathematician John Charles Fields.

Ludwik (Ludwig) Fleck (1896–1961). Polish medical doctor and biologist who developed the concept of thought collectives. This concept is important in the philosophy of science and the sociology of science. It helps to explain how scientific ideas change over time and is related to Thomas Kuhn's later notion of paradigm shift. Fleck felt that the development of scientific insights is not unidirectional and does not consist of just accumulating new pieces of information but also in overthrowing old ones.

Catherine Fosnot. U.S. mathematics educator. Her main research is on the application of current models of cognitive psychology to the teaching of mathematics. Fosnot and her colleagues designed realistic problem situations as the starting point of investigation, inviting learners "to mathematize" initially in their own informal ways. Classrooms are thereby turned into workshops with learners engaged in inquiry, subsequently proving and communicating their thinking to their peers.

Jean Baptiste Joseph Fourier (1768–1830). French mathematical physicist who established the partial differential equation governing heat diffusion and solved it by using infinite series of trigonometric functions.

Hans Freudenthal (1905–1990). Dutch mathematician who worked on the characters of the semisimple Lie groups between 1954 and 1956. Later he moved into the history of mathematics and mathematical education and had a great influence on the development of mathematics education research around the world.

Kurt Friedrichs (1901–1982). German-American mathematician who was coauthor with Richard Courant of the classic work *Supersonic Flow and Shock Waves*.

Ferdinand Georg Frobenius (1849–1917). German algebraist who was a professor at Berlin.

Dmitry Borisovich Fuchs (1939–). Russian-American mathematician who is a professor at the University of California, Davis, and prominent researcher in representations of infinite-dimensional Lie algebras. He was formerly a professor at Moscow State University and an instructor at the Jewish People's University.

Galileo Galilei (1564–1642). Italian scientist who formulated the basic law of falling bodies, which he verified by careful measurements. He constructed a telescope with which he studied lunar craters, discovered four moons revolving around Jupiter, and espoused the Copernican cause.

Howard Gardner (1943–). U.S. cognitive psychologist at Harvard Graduate School of Education. He is best known for his theory of multiple intelligences. In 1981 he was awarded a MacArthur Prize Fellowship.

Carl Friedrich Gauss (1777–1855). German mathematician who worked in a wide variety of fields in both mathematics and physics, including number theory, analysis, differential geometry, geodesy, magnetism, astronomy, and optics. His work has had an immense influence in many areas.

David Hillel Gelernter (1955–). U.S. computer scientist who is a professor of computer science at Yale University. In the 1980s, he made seminal contributions to the field of parallel computation. In 1993 Gelernter was critically injured opening a mail bomb sent by Theodore Kaczynski, who at that time was an unidentified but violent opponent of technology, dubbed by the press as "the Unabomber." Gelernter's right hand and eye were permanently damaged. He chronicled the ordeal in his 1997 book *Drawing Life: Surviving the Unabomber*.

Israel Moiseevich Gelfand (1913–2009). Russian mathematician who was one of the most prolific and original mathematicians of the 20th century. He published over 500 papers in mathematics, applied mathematics, and biology. He was deeply involved in education, and established a correspondence school that brought rich mathematical experiences to students all over the Soviet Union.

Paulus Gerdes. Mozambique professor of mathematics at the Eduardo Mondlane University and at the Universidade Pedagogica in

Mozambique for many years, serving as rector of the latter from 1989 to 1996. He is the director of the Ethnomathematics Research Centre and author of *Geometry from Africa*, *Women, Art and Geometry in Southern Africa*, and *Culture and the Awakening of Geometrical Thinking*.

Marie-Sophie Germain (1776–1831). French mathematician who made a major contribution to number theory, acoustics, and elasticity. She worked outside the established institutions that were closed to women.

Adele Gödel (1899–1981). Kurt Gödel's wife, an Austrian who for many years nurtured and protected him. When she suffered an incapacitating stroke, he cared for her devotedly until she required emergency surgery and was hospitalized for nearly 6 months. Then Gödel had to fend for himself, and his fear of poisoning led to self-starvation. He died on January 14, 1978. At her death she bequeathed the rights to Gödel's papers to the Institute for Advanced Study.

Kurt Gödel (1906–1978). Austrian philosopher and logician. His "incompleteness theorem" says that in any axiomatic mathematical system that includes the natural numbers, there are propositions that cannot be proved or disproved within the axioms of the system. At the Institute for Advanced Study he was a close friend of Albert Einstein.

Rebecca Goldstein (1950–). U.S. novelist and professor of philosophy. She has published six novels, a collection of stories, and two biographical studies—of mathematician Kurt Gödel and philosopher Baruch Spinoza.

Wayne Gould (1945–). Retired Hong Kong judge known for popularizing su do ku puzzles. Gould spent 6 years writing a computer program known as Pappocom Sudoku that mass-produces puzzles for the global market.

Evelyn Boyd Granville (1924–). Second African-American woman to earn a Ph.D. in mathematics. She was a student of Einar Hille at Yale. She worked as an applied mathematician for IBM and North American Aviation and taught at Fisk University in Nashville, at California State University in Los Angeles, and at Texas College in Tyler, Texas. Smith College, where she had earned her bachelor's degree, awarded her an honorary doctorate in 1989.

Mary Gray (1939–). U.S. professor at American University in Washington, D.C. whose focus is on statistics. For years her name was virtually synonymous with the Association for Women in Mathematics. She is also a human rights activist.

John W. Green (1943–). U.S. mathematician who was a student of Robert Lee Moore. Green was on the University of Oklahoma faculty for 15 years. He then obtained a Ph.D. in mathematical statistics from Texas A. & M. University and then taught at the University of Delaware for 5 years. He is currently employed by E. I. DuPont as a senior research biostatistician.

Alexandre Grothendieck (1928–). German-born but later stateless mathematician who worked in France. From 1959 to 1970 a whole new school of mathematics flourished under Grothendieck's leadership. His Séminaire de Géométrie Algébrique established the Insitut des Hautes Etudes Scientifiques (IHES) as a world center of algebraic geometry. He received the Fields Medal in 1966.

Jacques Salomon Hadamard (1865–1963). French mathematician who was a professor at the Sorbonne in Paris and proved the prime number theorem in 1896. This states that the number of primes $< n$ is asymptotic to $n/\log n$ as n tends to infinity.

Hans Hahn (1879–1934). Austrian mathematician remembered for the Hahn-Banach theorem. He made important contributions to the calculus of variations, developing ideas of Weierstrass.

Paul Halmos (1916–2006). Hungarian-American mathematician well known for graduate texts on mathematics dealing with finite-dimensional vector spaces, measure theory, ergodic theory, and Hilbert space. Many of these books were the first systematic presentations of their subjects in English.

Israel Halperin (1911–2007). Canadian mathematician and human rights activist. He was a graduate student of John von Neumann at Princeton University. In February 1946, he was arrested and accused of espionage in Canada in connection with the defection of Igor Gouzenko, a Soviet cipher clerk. After some arduous questioning and confinement lasting several weeks, Halperin was eventually cleared. He was elected a Fellow of the Royal Society of Canada in 1953 and won the Henry Marshall Tory Medal in 1967. He authored more than 100 academic papers, was awarded an honorary doctorate of laws from Queen's in 1989, and was made a Member of the Order of Canada for his humanitarian work and his mathematics.

George Bruce Halsted (1853–1922). U.S. mathematician. Eccentric and sometimes spectacular, Halsted became internationally known as a scholar, teacher, promoter, and popularizer of mathematics.

Richard Hamilton (1943–). U.S. mathematician who received a Ph.D. from Princeton University in 1966 under the direction of Robert Gunning. He is Professor of Mathematics at Columbia University and in 1996 was awarded the Oswald Veblen Prize of the American Mathematical Society.

Godfrey Harold Hardy (1877–1947). English mathematician who was a professor at Oxford and Cambridge, England. Hardy's interests covered many topics of pure mathematics: diophantine analysis, summation of divergent series, Fourier series, the Riemann zeta function, and the distribution of primes.

Jenny Harrison (1949–). U.S. professor of mathematics at the University of California at Berkeley. She developed a theory of quantum calculus that unifies an infinitesimal calculus with the classical theory of the smooth continuum. The methods apply to a class of domains called chainlets, which include soap films, fractals, graphs of L_1 functions, and charged particles, as well as smooth manifolds. Harrison's lawsuit, based on sex discrimination in the Berkeley mathematics department's tenure decision in 1987, attracted international attention.

Helmut Hasse (1898–1979). German mathematician who worked on algebraic number theory and was known for fundamental contributions to class field theory, the application of p-adic numbers to local class field theory and diophantine geometry (the Hasse principle) and to local zeta functions. He was Hermann Weyl's replacement at Göttingen in 1934. After the defeat of Germany he returned to Göttingen briefly but was excluded by the British authorities. After brief appointments in Berlin, in 1948 he settled permanently as a professor in Hamburg.

Eduard Helly (1884–1943). Austrian mathematician who worked on functional analysis and in 1912 proved the Hahn-Banach theorem, 15 years before Hahn and 20 years before Banach.

Ravenna Helson. U.S. professor at the University of California, Berkeley, adjunct professor emeritus of the Department of Psychology of Mills College, and director of the Mills Longitudinal Study, which she initiated. Long interested in gender issues, she works with students on data from the Mills Study, which has followed about 120 women for 40 years, from age 21 to age 61. The Mills Study examines long-term personality components, social influences on personality, and processes of growth and development.

Claudia Henrion. U.S. author of *Women in Mathematics: The Addition of Difference*. She lives in New Hampshire in the United States.

Wilhelm Heydorn (1873–1958). German Protestant theologian, medical practitioner, and teacher. From 1921 to 1923 he studied at the University of Hamburg and worked until 1926 as medical practitioner. From 1926 to 1928 he studied to become a primary school teacher and worked from 1928 to 1933 as an assistant teacher. From 1934 to 1939 he and his wife, Dagmar, took care of Alexandre Grothendieck.

David Hilbert (1862–1943). German mathematician who was the leader of the famous Göttingen school of pure and applied mathematics. Hilbert's work in geometry had the greatest influence in that area after Euclid. He made contributions in many areas of mathematics, physics, and logic.

Adolf Hurwitz (1859–1919). German mathematician who studied the genus of the Riemann surface and worked on how class number relations could be derived from modular equations.

Shokichi Iyanaga (1906–2006). Japanese mathematician who was a student of Teiji Takagi and contributed to algebraic number theory.

Fritz John (1910–1994). German-American mathematician who wrote classical papers on convexity, ill-posed problems, the numerical treatment of partial differential equations, quasi-isometry, and blow-up in nonlinear wave propagation.

Camille Jordan (1838–1922). French mathematician highly regarded for his work on algebra, group theory, and Galois theory.

Mark Kac (1914–1984). Polish-American mathematician who made major contributions to applied mathematics and probability theory. His sparkling juxtaposition of surprising phrases and constructions made his speech a delight to listen to.

Ted Kaczynski (1942–). U.S. mathematician (called "the Unabomber") who gave up a Berkeley professorship to go live in the woods in Montana.

Nicholas M. Katz (1943–). U.S. mathematician working in algebraic geometry, particularly on p-adic methods, monodromy and moduli problems, and number theory. He played a significant role as a sounding board for Andrew Wiles when Wiles was developing in secret his proof of Fermat's Last Theorem.

Felix Christian Klein (1849–1925). Very influential German mathematician. His Erlangen program defined a geometry as the properties of a space that are invariant under a given transformation group.

John R. Kline (1891–1955). U.S. professor of mathematics at the University of Pennsylvania from 1920 until his death in 1955. He was chair of the department from 1933 to 1954. He directed the Ph.D. theses of 13 students, including Dudley Weldon Woodard and W.W.S. Claytor, the second and third African-American mathematicians to earn a Ph.D.

Andrey Nikolaevich Kolmogorov (1903–1987). Russian mathemetics professor at the University of Moscow who became internationally famous for the rigorous development of measure-theoretic probability. He used this work to study the motion of the planets and the turbulent flow of air from a jet engine.

Sofia Vasilyevna Kovalevskaya (1850–1891). Russian mathematician, student, and friend of Karl Weierstrass, who made important contributions to the theory of differential equations.

Leopold Kronecker (1823–1891). German mathematician known for insisting that all mathematics must be based constructively on the natural numbers.

Ernst Kummer (1810–1893). German algebraist and number theorist famous for introducing "ideal numbers."

Kazimierz Kuratowski (1896–1980). Polish mathematician who worked on topology and set theory. He is best known for his theorem giving a necessary and sufficient condition for a graph to be planar.

Imre Lakatos (1922–1974). Hungarian-British philosopher of mathematics and science at the London School of Economics who was influenced by Karl Popper and George Polya.

Edmund Georg Hermann Landau (1877–1938). German mathematician who gave the first systematic presentation of analytic number theory and wrote important works on the theory of analytic functions.

Pierre-Simon, Marquis de Laplace (1749–1827). French mathematician and astronomer whose work was pivotal to the development of mathematical astronomy and statistics.

Jean Lave. U.S. social anthropologist whose research encompasses social practice in everyday mathematics, the study of apprenticeships, and learning communities. She is a professor at the University of California, Berkeley.

Anneli Lax (1922–1999). U.S. student of Richard Courant, professor at NYU, and editor of dozens of volumes of the New Mathematical Library.

Peter David Lax (1926–). Hungarian-American mathematician at the Courant Institute in New York. He is a leading researcher on partial differential equations and scattering theory and a winner of the Abel Prize (Norway) in 2005.

Solomon Lefschetz (1884–1972). U.S. mathematician who was the main source of work on the algebraic aspects of topology. He was a long-time chair at Princeton University.

Adrien-Marie Legendre (1752–1833). French mathematician who made important contributions to statistics, number theory, abstract algebra, and mathematical analysis.

Jean Leray (1906–1998). French mathematician who connected energy estimates for partial differential equations with fixed point theorems from algebraic topology in a highly original combination that cracked open the toughest problems. He was the first to adopt the modern viewpoint, where a function is thought of not as a complicated relation between two sets of variables but as a point in some infinite dimensional space.

Norman Levinson (1912–1975). U.S. Specialist in differential equations. He was considered the heart of mathematics at MIT, a man who combined high creative intellect with human compassion and dedication to science.

Roy Lisker (1938–). U.S. mathematician, novelist, musician, publisher, and social critic. Lisker publishes *Ferment* magazine and is the proprietor of the Ferment Press.

John Edensor Littlewood (1885–1977). Outstanding English mathematician. He was Hardy's collaborator and worked on the theory of series, the Riemann zeta function, inequalities, and the theory of functions.

Nikolai Ivanovich Lobachevsky (1792–1856). Russian mathematician, at the University of Kazan, who in 1829 published his non-Euclidean geometry, the first account of the subject to appear in print.

Lee Lorch (1915–). U.S. mathematician and early civil rights activist who is currently professor emeritus at York University in Toronto, Canada. As a teacher at Black universities such as Fisk and Philander Smith, Lorch encouraged students, including Black women, to pursue graduate study in mathematics. Two colleges that fired him, Fisk University and City University, later awarded him honorary degrees. He was also honored by the U.S. National Academy of Sciences in 1990 and by Spelman College.

Edith Luchins (1921–2002). U.S. mathematician who taught at the Rensselaer Polytechnic Institute from 1962 until 1992. She was the first woman to be appointed a full professor at Rensselaer. Luchins' research focused on mathematics and psychology. She worked on mathematical models of order effects in information processing; on gender differences in cognitive processes and their implications for teaching and learning mathematics; and on the roles of heuristics and algorithms in mathematical problem solving.

George Whitelaw Mackey (1916–2006). U.S. mathematician who joined the Harvard University mathematics department in 1943, was appointed Landon T. Clay Professor of Mathematics and Theoretical Science in 1969, and remained there until he retired in 1985. Mackey's main research was in representation theory, ergodic theory, and related parts of functional analysis. Mackey did significant work on the duality theory of locally convex spaces, which provided tools for subsequent work in this area, including Alexandre Grothendieck's work on topological tensor products.

MacTutor History of Mathematics. An extensive, searchable online archive of persons and concepts, written and edited by J. J. O'Connor and E. F. Robertson at the University of St. Andrews.

Wilhelm Magnus (1907–1990). German-American mathematician. In addition to his main research on group theory and special functions, he worked on problems in mathematical physics, including electromagnetic theory and applications of the wave equation. He was an outstanding supervisor of doctoral students at the Courant Institute, supervising 61 dissertations during his career.

Vivienne Malone-Mayes (1932–1995). U.S. professor who earned a B.A. (1952) and M.A. (1954) in mathematics at Fisk University. Dr. Mayes became an outstanding teacher as the first Black faculty member at Baylor University, which had rejected her as a student with an explicit antiblack policy 5 years earlier.

Benoit Mandelbrot (1924–). French-American mathematician who is a major contributor to the growing field of fractal geometry. He showed that fractals occur in many different places, in mathematics and elsewhere in nature.

José Luis Massera (1915–2002). Uruguayan mathematician. A theorem that bears his name solves the problem of the stability of nonlinear differential equations in terms of the Lyapunov exponents. Massera's political activity resulted in his arrest on October 22, 1975. A broad and vigorous international campaign constantly demanded his release, and he regained freedom in March 1983.

Stanislaw Mazur (1905–1981). Polish mathematician who made important contributions to geometrical methods in linear and nonlinear functional analysis and to the study of Banach algebras. Stan Ulam recounts how Mazur gave the first examples of infinite games in the Scottish Café in Lvov.

John Milnor (1931–). U.S. mathematician. His most remarkable achievement, which played a major role in his winning the Fields Medal, was his proof that a seven-dimensional sphere can have several differential structures. This work opened up the new field of differential topology.

Hermann Minkowski (1864–1909). German mathematician who was a friend and collaborator of Hilbert. He developed a new view of space and time and laid the mathematical foundation of the theory of relativity.

Magnus Gösta Mittag-Leffler (1846–1927). Swedish mathematician who was a student of Weierstrass. He worked on the general theory of functions. His best known work concerned the analytic representation of a one-valued function.

Edwin Evariste Moise (1918–1998). U.S. mathematician. He was a topologist, a student of Robert Lee Moore. He helped decipher German and Japanese military signals in World War II. At Michigan he did important work on 3-manifolds, culminating in his proof that every 3-manifold can be triangulated.

Gaspard Monge (1746–1818). French mathematician considered the father of differential geometry because of his theory of curvature of a surface in 3-space.

Calvin C. Moore. U.S. functional analyst who has had a long career as professor and administrator at the University of California, Berkeley.

Robert Lee Moore (1882–1974). U.S. mathematician known for his work on general topology and the Moore method of teaching university mathematics.

Cathleen Morawetz (1923–). U.S. mathematician who was the second female president of the American Mathematical Society in 1995–1996. In 1998 she was awarded the National Medal of Science for pioneering advances in partial differential equations and wave propagation resulting in applications to aerodynamics, acoustics, and optics.

Louis Joel Mordell (1888–1972). British-American mathematician best known for his investigations of equations of the form $y^2 = x^3 + k$, which had been studied by Fermat. His conjecture about rational points on algebraic curves, which was proved in 1983 by Gerd Faltings, was an important ingredient in Andrew Wiles' proof of Fermat's Last Theorem.

Oskar Morgenstern (1902–1977). Austrian-American economist who, working with John von Neumann, helped found the mathematical field of game theory.

Harold Calvin Marston Morse (1892–1977). U.S. mathematician who developed variational theory in the large with applications to equilibrium problems in mathematical physics, a theory that is now called Morse theory and forms a vital role in the mathematics of global analysis.

Jürgen Moser (1928–1999). German mathematician who worked at the Courant Institute in New York and at the ETH in Geneva. He specialized in ordinary differential equations, partial differential equations, spectral theory, celestial mechanics, and stability theory. The leading theme of virtually all of Moser's work in dynamics is the search for elements of stable behavior in dynamical systems with respect to either initial conditions or perturbations of the system.

Robert Parris Moses (1935–). U.S. Harvard-trained educator (known as Bob Moses) who was a leader in the civil rights movement and later founded the nationwide U.S. Algebra Project.

Gilbert Murray (1948–1995). Former president of the California Forestry Association, a timber industry lobbying group. In 1995 he was killed by a bomb mailed by "the Unabomber," Ted Kaczynski. The bomb was addressed to the CFA's previous president, Bill Dennison.

John Nash (1928–). U.S. mathematician. In 1949, while studying for his doctorate, he wrote a paper which 45 years later won a Nobel Prize for economics. P. Ordeshook wrote: "The concept of a Nash equilibrium n-tuple is perhaps the most important idea in noncooperative game theory."

Rolf Nevanlinna (1895–1980). Finnish mathematician whose most important work was on harmonic measure, which he invented. He also developed the theory of value distribution named after him. The main results of Nevanlinna theory appeared in 1925 in a 100-page paper that Weyl called "one of the few great mathematical events of our century."

Isaac Newton (1643–1727). English mathematician and physicist who laid the foundation for differential and integral calculus. His work on optics and gravitation makes him one of the greatest scientists the world has known.

Jerzy Neyman (1894–1981). Polish-American statistician who put forward the theory of confidence intervals, which plays a central role in statistical theory and data analysis. His contribution to the theory of contagious distributions is still of great utility in the interpretation of biological data.

Nel Noddings (1929–). American feminist and philosopher best known for her work on the philosophy of education, educational theory, and the ethics of care.

Emmy Noether (1882–1935). German mathematician who was the principal founder of modern abstract algebra. She is particularly known for her study of chain conditions on ideals of rings. To this day, she is an inspiration to women mathematicians.

Sergei Novikov (1938–). Russian mathematician and Fields medalist. Both his parents were famous mathematicians. Novikov studied at the Faculty of Mathematics and Mechanics of Moscow University (1955–1960) and has worked there since 1964 in the department of differential geometry. His work has played an important part in building a bridge between modern mathematics and theoretical physics.

King Oscar II (1829–1907). Successor to his brother, Carl IV, as King of Norway and Sweden in 1872. He married Sophia of Nassau, and their eldest son, Gustav, became King of Sweden. Oscar II was the last king of the Norwegian-Swedish union. Oscar supported mathematical research in Sweden, and an important prize was named after him.

Blaise Pascal (1623–1662). Influential French mathematician and philosopher who contributed to many areas of mathematics. He worked on conic sections and projective geometry, and in correspondence with Fermat he laid the foundations for the theory of probability.

John Allen Paulos (1945–). U.S. professor of mathematics at Temple University in Philadelphia who has written on the importance of mathematical literacy and the mathematical basis of humor.

Sir Roger Penrose (1931–). English mathematical physicist renowned for contributions to general relativity and cosmology. He is also a recreational mathematician and philosopher. Penrose proved that, under certain conditions, in a gravitational collapse space-time cannot be continued and classical general relativity breaks down. He looked for a unified theory combining relativity and quantum theory since quantum effects become dominant at the singularity. One of Penrose's major breakthroughs was his twistor theory, an attempt to unite relativity and quantum theory.

Grigori Yakovlevich Perelman (1966–). Russian mathematician who has made landmark contributions to Riemannian geometry and geometric topology. In particular, he proved Thurston's geometrization conjecture. This solves in the affirmative the famous Poincaré conjecture posed in 1904 and regarded as one of the most important and difficult open problems in mathematics. In August 2006, Perelman was awarded the Fields Medal for "his contributions to geometry and his revolutionary insights into the analytical and geometric structure of the Ricci flow." However, he declined to accept the award or to appear at the congress.

Rózsa Péter (1905–1977). Hungarian mathematician best known for her work on recursion theory. She attended Eötvös Loránd

University, where she received a Ph.D. in 1935. During Miklós Horthy's collaboration with Nazi Germany, she was forbidden to teach because of her Jewish origin. After the war she published her key work, *Recursive Functions*. She taught at her alma mater from 1955 until her retirement in 1975.

Ivan Georgievich Petrovsky (1901–1973). Russian mathematician in the field of partial differential equations. He greatly contributed to the solution of Hilbert's 19th and 16th problems. He also worked on boundary value problems, probability, and topology of algebraic curves and surfaces. Among his students were Olga Ladyzhenskaya and Olga Oleinik. Petrovsky was the rector of Moscow State University (1951–1973). and the head of the International Congress of Mathematicians (Moscow, 1966). He regarded his rectorship of Moscow State University as the most important thing in his life, even more important than his mathematical research.

Wolodymyr V. Petryshyn (1929–). U.S. mathematician whose major results include the development of the theory of iterative and projective methods for the constructive solution of linear and nonlinear abstract and differential equations.

Jean Piaget (1896–1980). Swiss developmental psychologist well known for his work studying children, his theory of cognitive development, and his theory of knowledge acquisition called "genetic epistemology." He was one of the major figures of 20th century psychology. In 1955 he created the International Centre for Genetic Epistemology in Geneva and directed it until 1980.

Vera Pless (1931–). U.S. mathematician who worked at the Air Force Cambridge Research Laboratory from 1963 to 1972, where she became a leading expert on coding theory. In 1963 she published power moment identities on weight distributions in error-correcting codes. These are used to determine the complete weight distributions of several quadratic residue codes.

Jules Henri Poincaré (1854–1912). French mathematician who was one of the 20th century giants in this field. He was the originator of algebraic topology and of the theory of analytic functions of several complex variables.

Siméon-Denis Poisson (1781–1840). French mathematician, geometer, and physicist.

George Polya (1887–1985). Hungarian-American mathematician who worked on probability, analysis, number theory, geometry, combinatorics, and mathematical physics. His writings on heuristics and problems solving are very influential.

Jean-Victor Poncelet (1788–1867). French mathematician who was one of the founders of modern projective geometry. His development of the pole and polar lines associated with conics led to the principle of duality.

Lev Pontryagin (1908–1988). Russian mathematician who constructed a general theory of characters for commutative topological groups. He used this theory of characters to prove that any Abelian locally Euclidean topological group can be given the structure of a Lie group (Hilbert's 5th problem for the case of Abelian groups).

Karl Popper (1902–1994). Austrian who was a major philosopher of science and opposed all forms of skepticism, conventionalism, and relativism. In human affairs he was a committed advocate and defender of an "open society."

Marian Boykan Pour-El (?–2009). U.S. logician and mathematician who was the author of many articles on logic and its applications in mathematics and physics. She specialized in computability and noncomputability of mathematical and physical theories.

Tibor Radó (1895–1965). Hungarian-American mathematician. The Plateau problem was studied by Plateau, Weierstrass, Riemann, and Schwarz but was finally solved by Douglas and Radó. He used conformal mappings of polyhedra, applying a limit theorem to certain approximations to obtain the minimal surface required. The solution did not exclude the possibility that the minimal surface might contain a singularity. In fact, it never does contain a singularity; this was shown for the first time by Osserman in 1970.

George Yuri Rainich (1886–1968). Russian-American mathematical physicist who was well-known in the early 20th century. Rainich's research centered on general relativity and early work toward a unified field theory.

Srinivasa Aiyangar Ramanujan (1887–1920). Indian mathematician who became well known through his close collaboration with G. H. Hardy. Ramanujan made substantial contributions to the analytical theory of numbers and worked on elliptic functions, continued fractions, and infinite series.

Lord Rayleigh (John William Strutt) (1842–1919). English physicist whose theory of scattering, published in 1871, was the first correct explanation of why the sky is blue. The first volume of his major text, *The Theory of Sound*, on the mechanics of a vibrating medium, was published in 1877, and the second volume, on acoustic wave propagation, was published the following year.

Constance Bowman Reid (1918–). U.S. author of several biographies of mathematicians and popular books about mathematics. She is the sister of mathematician Julia Robinson.

Alfred Rényi (1921–1970). Hungarian mathematician. He obtained a Ph.D. at Szeged under F. Riesz, was taught by Fejér at Budapest, and then went to Russia to work with Linnik on the theory of numbers, in particular on the Goldbach conjecture. He discovered methods described by Turán as among the strongest methods of analytical number theory. After returning to Hungary he worked on probability. He published joint work with Erdős on random graphs and also considered random space filling curves. He is remembered for proving that every even integer is the sum of a prime and an almost prime number (one with only two prime factors).

Bernhard Riemann (1826–1866). German mathematician who was a student of Gauss and a professor at Göttingen. His ideas concerning the geometry of space had a profound effect on the development of modern theoretical physics. He clarified the notion of an integral by defining what we now call the Riemann integral.

Frigyes Riesz (1880–1956). Hungarian mathematician who was a founder of functional analysis and whose work has many important applications in physics. He was the leader of the mathematical school at Szeged.

Herbert Ellis Robbins (1915– 2001). U.S. mathematician and statistician who did research on topology, measure theory, statistics, and a variety of other fields. He was the coauthor, with Richard Courant, of *What Is Mathematics?*, a popularization that is still (as of 2009) in print. The Robbins lemma, used in empirical Bayes methods, is named after him.

Abraham Robinson (1918–1974). Israeli-American mathematician whose most famous invention was nonstandard analysis which he introduced in 1961. A nonstandard model for the system of real numbers has the feature of being a non-Archimedean totally ordered field that contains a copy of the real number system.

Julia Hall Bowman Robinson (1919–1985). U.S. mathematician who worked on computability, decision problems, and nonstandard models of arithmetic. The recipient of many honors, Robinson was the first female mathematician member of the National Academy of Sciences and the first female president of the American Mathematical Society. She also received a McArthur grant.

Raphael Robinson (1911–1995). U.S. professor of mathematics at Berkeley, who worked on complex analysis, logic, set theory, geometry, number theory, and combinatorics.

Judith Roitman (1945–). U.S. mathematician who is currently a professor at the University of Kansas. She specializes in applications of set theory to topology and Boolean algebra.

Gian-Carlo Rota (1932–1999). Italian-American professor at MIT in applied mathematics and philosophy. He began as a functional analyst and moved on to combinatorics, where he became a leading national and international figure. He was also a person of great cultural and literary attainment. He was a mathematical gourmet.

Henry Roth (1906–1995). U.S. author whose best-known work is *Call It Sleep* (1934), a classic in Jewish-American literature.

Mary Ellen (Estill) Rudin (1924–). U.S. mathematician. In 1971 the University of Wisconsin promoted her from lecturer to full professor. "Nobody even asked me if I wanted to be a professor. I was simply presented with this full professorship." Her research has been in set-theoretic topology, especially using axiomatic set theory. Methods from general topology unexpectedly solved several difficult problems in functional analysis and in geometric and algebraic topology. Two developments revolutionized the field: the creation of infinite-dimensional topology and set-theoretic topology. It has been mainly due to Dick Anderson and Mary Ellen Rudin that these fields have dominated general topology ever since.

Carle David Tolmé Runge (1856–1927). German mathematician who worked on a procedure for the numerical solution of algebraic equations and later studied the wavelengths of the spectral lines of elements.

Bertrand Arthur William Russell (1872–1970). English mathematician and philosopher who published a vast number of books on logic, theory of knowledge, and many other topics. His best known work was *Principia Mathematica*.

Oliver Sacks (1933–). U.S.-based British neurologist who has written popular books about brain pathologies in his patients, the most famous of which is *Awakenings*.

Pierre Samuel (1921–2009). French mathematician who worked with Alexandre Grothendieck's environmentalist movement and was a cofounder of the French Green Party. He was known for work in commutative algebra and its applications to algebraic geometry. The two-volume work *Commutative Algebra* that he wrote with Oscar Zariski is a classic.

Jane Cronin Scanlon (1922–). U.S. mathematician who began her college studies in physics but later become more interested in the abstract nature of mathematics. She received a Ph.D. in mathematics in 1949 from the University of Michigan. Her research interests are in mathematical biology and partial differential equations.

Alice Schafer (1915–2009). U.S. mathematician. She lists two special-izations: abstract algebra (group theory) and women in mathemat-ics. Through her work as a mathematics educator, she became a champion for full participation of women in mathematics. From 1962 to 1980 she was professor and head of the mathematics de-partment at Wellesley College.

Laurent Schwartz (1915–2002). French mathematician whose paper on the theory of distributions is one of the classical mathematical papers of our time. Later he worked on stochastic differential cal-culus. He campaigned against American involvement in Vietnam, the Soviet invasion of Afghanistan, and the Russian war against Chechnya. He also was an avid collector of butterflies, with over 20,000 specimens. Schwartz wrote in his autobiography, "To dis-cover something in mathematics is to overcome an inhibition and a tradition. You cannot move forward if you are not subversive."

Atle Selberg (1917–2007). Norwegian mathematician who was awarded a Fields Medal for his work on generalizations of the sieve methods of Viggo Brun, and for his major work on the zeros of the Riemann zeta function, where he proved that a positive pro-portion of its zeros satisfy the Riemann hypothesis. Selberg is also well known for his elementary proof of the prime number theorem, which says that the number of primes $\leq n$ is asymptotic to $n/\log n$ as n tends to infinity.

Jean-Pierre Serre (1926–). French mathematician who is one of the leading mathematicians of the 20th century, active in algebraic ge-ometry, number theory, and topology. His early work was on spec-tral sequences. The Serre spectral sequence provided a tool to work effectively with the homology of fiberings. For this work on spec-tral sequences and his work developing complex variable theory in terms of sheaves, Serre was awarded a Fields Medal at the Interna-tional Congress of Mathematicians in 1954. Serre's theorem led to rapid progress not only in homotopy theory but also in algebraic topology and homological algebra in general.

Igor Rostislavovich Shafarevich (1923–). Russian mathematician who was the founder of the major school of algebraic number theory and algebraic geometry in the former Soviet Union. He made ma-jor contributions to the inverse problem of Galois theory as well as to class field theory, settling some long-outstanding conjectures. More recently he has made important advances in algebraic geometry. He was an important dissident figure under the Soviet regime, a public supporter of Andrei Sakharov's human rights committee from 1970. In 1972, he joined the group of dissidents led by Solzhenitsyn. As a consequence, he was dismissed from the University of Moscow in 1975. In 1989 he published a book, *Russophobia,* which contained surprising anti-Semitic slanders.

Allen Lowell Shields (1927–1989). U.S. mathematician who worked on a wide range of mathematical topics including measure theory, complex functions, functional analysis, and operator theory.

Isadore Singer (1924–). U.S. mathematician famous for his deep and spectacular work on geometry, analysis, and topology. Singer's five papers with Michael F. Atiyah on the index theorem for elliptic operators and his three papers with Atiyah and V. K. Patodi on the index theorem for manifolds with boundary are among the great classics of global analysis. They have spawned many developments in differential geometry, differential topology, and analysis.

Steve Smale (1930–). U.S. mathematician who began his career as an instructor at the University of Chicago. In 1958 he astounded the mathematical world with a proof of a sphere eversion. He then cemented his reputation with a proof of the Poincaré conjecture for all dimensions greater than or equal to 5. He later generalized the ideas in a 107-page paper that established the *h*-cobordism theorem. After making great strides in topology, he then turned to the study of dynamical systems. His first contribution was the Smale horseshoe, which jump-started significant research in dynamical systems. He also outlined a research program carried out by many others. Smale is also known for injecting Morse theory into mathematical economics, as well as recent explorations of various theories of computation. He is also known as world-class collector and photographer of mineral specimens.

Alexei B. Sossinsky (1937–). Russian senior researcher in the laboratory of mathematical methods of the Institute for Problems in Mechanics at the Russian Academy of Sciences, and professor at the Higher Mathematics College of the Independent University of Moscow.

Julian Cecil Stanley (1918–2005). U.S. psychologist, educator, and advocate of accelerated education for academically gifted children. He founded the Johns Hopkins University Center for Talented Youth (CTY), as well as a related research project, the Study of Mathematically Precocious Youth, which in 2005 was renamed the Julian C. Stanley Study of Exceptional Talent (SET). Stanley was also widely known for his book, coauthored with Donald Campbell, *Experimental and Quasi-experimental Designs for Research*.

Clarence F. Stephens (1917–). U.S. mathematician who graduated from Johnson C. Smith University in 1938 with a B.S. degree in mathematics. He received an M.S. (1939) and a Ph.D. (1943) from the University of Michigan. In 1947 he joined the mathematics faculty at Morgan State University. Stephens remained at Morgan State until 1962, at which time he accepted an appointment as professor of mathematics at SUNY Geneseo. In 1969 he left Geneseo to join the mathematics faculty at SUNY Potsdam, where he served as chair of the mathematics department until his retirement in 1987. During Stephens' tenure at SUNY Potsdam, the department became nationally known as a model of teaching excellence in mathematics.

Marshall Harvey Stone (1903–1989). U.S. mathematician who was a noted chairman at the University of Chicago. He is best known for the Stone-Weierstrass theorem on uniform approximation of continuous functions by polynomials.

Ted Streleski (1936–). U.S. graduate student in mathematics at Stanford University who murdered professor Karel de Leeuw with a ball peen hammer. Streleski felt the murder was justifiable because de Leeuw had withheld departmental awards from him and demeaned him in front of his peers. Streleski had been pursuing a doctorate in the mathematics department for 19 years. He served 7 years in prison and was eligible for parole on three occasions but turned it down because the conditions of parole required him to not set foot on the Stanford campus. Upon release he said, "I have no intention of killing again. On the other hand, I cannot predict the future."

Steven H. Strogatz (1959–). U.S. mathematician and Jacob Gould Schurman Professor of Applied Mathematics at Cornell University who is known for contributions to the study of synchronization in dynamical systems and for work in a variety of areas of applied mathematics, including mathematical biology and complex network theory.

Dirk Jan Struik (1894–2000). Dutch-American mathematician and Marxist theoretician who spent most of his life in the United States as a professor at MIT. He was a specialist in differential geometry.

Beauregard Stubblefield (1923–). U.S. mathematician who from 1952 to 1956 was head of the department of mathematics of the University of Liberia at Monrovia. From 1959 to 1960 he served as lecturer and National Science Foundation Post-Doctoral Fellow at the University of Michigan in Ann Arbor, Michigan. Since then he has been a faculty member at Stevens Institute of Technology in Hoboken, New Jersey, at Oakland University in Rochester, Michigan, at Texas Southern University, and at Appalachian State University in Boone, North Carolina.

Bella Abramovna Subbotovskaya (1938–1982). Russian mathematician who was the founder of the Jewish People's University in Moscow from 1978 to 1983.

James Joseph Sylvester (1814–1897). English mathematician who did important work on matrix theory. In 1851 Sylvester discovered the discriminant of a cubic equation and first used the name "discriminant" for equations of higher order. He founded the first U.S. graduate department of mathematics at Johns Hopkins, and the *American Journal of Mathematics.*

John Lighton Synge (1897–1995). Irish-Canadian professor who made outstanding contributions to classical mechanics, geometrical mechanics and geometrical optics, gas dynamics, hydrodynamics, elasticity, electrical networks, mathematical methods, differential geometry and, above all, Einstein's theory of relativity.

Gábor Szegö (1895–1985). Hungarian-American mathematician who worked on extremal problems and Toeplitz matrices. He was a professor and chair of mathematics at Stanford University and a longtime collaborator of George Polya.

Teiji Takagi (1875–1960). Japanese mathematician who was a student of Hilbert. He worked on class field theory, building on Heinrich Weber's work.

Yutaka Taniyama (1927–1958). Japanese mathematician who posed two problems at the Symposium on Algebraic Number Theory held in Tokyo in 1955 that form the basis of the Shimura-Taniyama conjecture: "Every elliptic curve defined over the *rational* field is a factor of the Jacobian of a modular function field." This conjecture was a major factor in Wiles' proof of Fermat's Last Theorem.

Olga Taussky-Todd (1906–1995). Distinguished and prolific Austrian-American mathematician who wrote about 300 papers. Her best known and most influential work was in matrix theory, and she also made important contributions to number theory.

Jean E. Taylor (1944–). U.S. mathematician who went to the University of California to study chemistry but audited S. S. Chern's class in differential geometry, which inspired her. With his help, she transferred to mathematics She then earned a M.Sc. in mathematics at Warwick and a Ph.D. at Princeton. In 1973 she went to Rutgers and rose to the rank of professor. In 2002 she retired and settled into the Courant Institute. Her research has been primarily in the field of geometric measure theory applied to problems of optimal shapes of crystals, both in equilibrium and otherwise.

Richard Taylor (1962–). English mathematician who received his Ph.D. from Princeton University in 1988. From 1995 to 1996 he held the Savilian Chair of Geometry at Oxford University and is currently the Herschel Smith Professor of Mathematics at Harvard University. One of the two papers containing the published proof of Fermat's Last Theorem is a joint work of Taylor and Andrew Wiles. In subsequent work, Taylor (along with Michael Harris) proved the local Langlands conjectures for GL(n) over a number field. Taylor, along with Christophe Breuil, Brian Conrad, and Fred Diamond, completed the proof of the Taniyama-Shimura conjecture.

William Thurston (1946–). U.S. professor at Cornell whose ideas revolutionized the study of topology in four dimensions and brought about a new interplay between analysis, topology, and geometry. This method is a new level of geometrical analysis—in the sense of powerful geometrical estimation on the one hand and spatial visualization and imagination on the other.

John Todd (1911–2007). English mathematician who spent 10 years in Washington at the National Applied Mathematical Laboratories, developing high-speed computer programming and becoming a world leader in numerical analysis and numerical algebra. In 1957 Todd and Olga Taussky moved to Cal Tech, where Todd developed the first undergraduate courses in numerical analysis and numerical algebra, prerequisites to learning computing.

Pál Turán (1910–1976). Hungarian mathematician. The most important and original of Turán's results are in his power sum method and its applications. They led to interesting deep problems of a completely new type; they have quite unexpectedly surprising consequences in many branches of mathematics—differential equations, numerical algebra, and various branches of function theory.

Yoshisuke Ueda (1936–). Japanese mathematician and engineer who discovered chaos in 1961 in the course of studying certain differential equations with the use of analog computers.

Stanislaw Marcin Ulam (1909–1984). Polish-American mathematician who was long associated with the Los Alamos National Laboratory, where he solved mathematically the problem of how to initiate fusion in the hydrogen bomb. He also promoted the Monte Carlo method widely used in solving mathematical problems using statistical sampling.

Kristin Umland. U.S. professor of mathematics at the University of New Mexico. Her research focuses on the cognitive aspects of learning mathematics.

Georges Valiron (1884–1955). French mathematician who was secretary of the assembly meeting of the International Mathematical Union in 1932.

Bartel Leendert van der Waerden (1903–1996). Dutch algebraist and historian who was the author of a very influential textbook on abstract algebra.

Harry Schultz Vandiver (1882–1973). U.S. mathematician. At an early age he went to work as a customs house broker for his father's firm. He never graduated from high school. When he was 18 he began to solve number theory problems in the *American Mathematical Monthly*. G. D. Birkhoff persuaded Vandiver to accept a post at Cornell University in 1919. In 1924 he moved to the University of Texas, being promoted to full professor in 1925 and then to distinguished professor of applied mathematics and astronomy in 1947. He retired in 1966 at the age of 84. He continued to work on extending Kummer's methods of studying Fermat's Last Theorem for increasingly large exponents. With hand calculations and help from his students, he had shown the result to be true for all *n* up to 600. In 1952 he was able to implement his methods on early computers and to prove the theorem true for all primes less than 2000.

Srinivasa Varadhan (1940–). Indian-American probabilist, professor at the Courant Institute at NYU, and winner of the Abel Prize in 2007 for fundamental contributions to probability theory and in particular for creating a unified theory of large deviation.

Oswald Veblen (1880–1960). U.S. mathematician and Princeton professor who made important contributions to projective and differential geometry, and topology.

William Spencer Vickrey (1914–1996). Columbia University economics professor whose Nobel Memorial Prize in Economics was announced just 3 days before he died. He authored the seminal 1961 *Journal of Finance* paper, "Counterspeculation, Auctions and Competitive Sealed Tenders," which was the first instance of an economist using the tools of game theory to understand auctions.

Theodor von Kármán (1881–1963). Hungarian-American mathematician who founded the U.S. Institute of Aeronautical Sciences, continuing his research on fluid mechanics, turbulence theory, and supersonic flight. He studied applications of mathematics to engineering, aircraft structures, and soil erosion.

Niels Fabian Helge von Koch (1870–1924). Swedish mathematician who gave his name to the famous fractal known as the Koch snowflake. He was born into a family of Swedish nobility. His father Richert Vogt von Koch (1838–1913) was a lieutenant colonel in the Royal Horse Guards of Sweden. Von Koch wrote several papers on number theory. One of his results was a 1901 theorem proving that the Riemann hypothesis is equivalent to a strengthened form of the prime number theorem.

John von Neumann (1903–1957). Hungarian-American professor at the Institute for Advanced Study in Princeton. He built a solid framework for quantum mechanics. He also worked in game theory, studied what are now called von Neumann algebras, and was one of the pioneers of computer science.

Lev Semenovich Vygotsky (1896–1934). Soviet developmental psychologist and the founder of cultural-historical psychology. He graduated from Moscow State University in 1917. While at the Institute of Psychology in Moscow (1924–1934), he worked extensively on ideas about cognitive development, particularly the relationship between language and thinking. His writings emphasized the roles of historical, cultural, and social factors in cognition and language. Vygotsky died of tuberculosis in 1934, leaving a wealth of work that is still being explored.

Janice Anita Brown Walker (1949–). U.S. mathematician who earned her Ph.D. at Michigan in 1982. The atmosphere in the mathematics department, the support of many faculty members, and the camaraderie among the students made her time at Michigan rewarding, stimulating, and comfortable. Since 1992 she has been chair of the Mathematics and Computer Science Department at Xavier University in Cincinnati, Ohio.

Valerie Walkerdine. English specialist on gender and class and its impact on mathematical development. She teaches at the School of Social Science at the University of Cardiff in Wales.

Karl Theodor Wilhelm Weierstrass (1815–1897). German mathematician who was a very influential professor at Berlin. He was best known for his construction of the theory of complex functions by means of power series.

André Weil (1906–1998). French mathematician and a founding member of Bourbaki. He started the rapid advance of algebraic geometry and number theory by laying the foundations for abstract algebraic geometry and the modern theory of Abelian varieties. His work on algebraic curves has influenced a wide variety of areas, including some outside mathematics such as elementary particle physics and string theory.

Tilla Milnor Weinstein (d. 2002). U.S. geometer specializing in Lorentz surfaces. She was a professor and then chair of mathematics at Douglass College, Rutgers University.

Anna Pell Wheeler (1883–1966). U.S. mathematician who was the first female mathematician to address the American Mathematical Society. She fostered women's participation in mathematics. Teaching longest at Bryn Mawr (1918–1948), where she also chaired the mathematics department, she specialized in integral equations. She was also an avid bird watcher and wildflower enthusiast.

Sylvia Wiegand (1945–). U.S. mathematician. Several members of her family were mathematicians. She wrote about her grandparents, Grace Chisholm Young and William Henry Young. Her interest in mathematics developed at a young age in response to the mathematical puzzles posed by her father, Laurence Chisholm Young, a mathematics professor at the University of Wisconsin. She received a Ph.D. in 1972 with a thesis in commutative algebra. She has co-authored many research papers with her husband, Roger Wiegand. They both hold appointments at the University of Nebraska.

Leo Wiener (1862–1939). U.S. professor of languages. Norbert Wiener's father, Leo, attended medical school at the University of Warsaw but then went to Berlin where he began training as an engineer. He gave this up to emigrate to America. Arriving in New Orleans in 1880, Leo tried various jobs in factories and farms before becoming a school teacher in Kansas City. He progressed to the post of professor of modern languages at the University of Missouri and then left Missouri for Boston. He was appointed an instructor in Slavic languages at Harvard, but this did not pay enough to provide for his family and he kept other positions to augment his salary. He remained at Harvard and eventually was promoted to professor.

Norbert Wiener (1894–1964). U.S. mathematics professor at MIT. Best known for his very influential work on Brownian motion. He introduced a measure on the space of one-dimensional paths which brings in probability concepts in a natural way. From 1923 he investigated Dirichlet's problem, producing work that had a major influence on potential theory.

Eugene Wigner (1902–1995). Hungarian-American physicist whose investigations of symmetry principles in physics are important far beyond nuclear physics proper. His methods and results have become an indispensable guide for the interpretation of the rich and complicated picture that has emerged from recent years' experimental research on elementary particles.

Raymond L. Wilder (1896–1982). U.S. mathematician who received a Ph.D. from the University of Texas in 1923. The bulk of his academic career was at the University of Michigan. Wilder was one of the developers of algebraic topology. He was the author of several books on the philosophy of mathematics and the culture of mathematics.

Andrew John Wiles (1953–). English number theorist and professor at Princeton University. In 1995 Wiles proved Fermat's Last Theorem, which states that if $a^n + b^n = c^n$, where a, b, c, and n are all positive integers, then n is less than or equal to 2.

Ellen Winner. U.S. professor of psychology at Boston College. Her work in the visual arts has focused on children's sensitivity to aesthetic aspects of works of art, such as line quality, expression, and composition. Her most recent book explores misconceptions about children who are extremely gifted and makes a set of recommendations about how such children should be educated.

Melanie Matchett Wood (1981–). U.S. mathematician who, while a high school student in Indianapolis, became the first female to make the U.S. International Mathematical Olympiad Team, receiving silver medals in the 1998 and 1999 competitions. She attended Duke University, where she was named a Putnam Fellow. During the 2003–2004 year she studied at Cambridge University, where she won the Morgan Prize for work in Belyi-extending maps and in P-orderings. She was named the deputy leader of the U.S. team that finished second overall at the 2005 International Mathematical Olympiad.

Dudley Weldon Woodard (1881–1965). African-American mathematician who attended Wilberforce College in Ohio, receiving a bachelor degree (A.B.) in mathematics in 1903. He then received a B.S. degree in 1906 and an M.S. degree in mathematics from the University of Chicago in 1907. From 1907 to 1914, he taught mathematics at Tuskegee Institute and then moved to join the Wilberforce faculty from 1914 to 1920. In 1921 he joined the mathematics faculty at Howard University. He received a Ph.D. in mathematics in 1928 at the University of Pennsylvania. While at Howard, he was also selected dean of the College of Arts and Sciences.

John Worrall (1946–). English philosopher who studied for a Ph.D. under Lakatos at the London School of Economics, developing the latter's methodology of research programs and testing it against a detailed case history from 19th century physics. Worrall was appointed to a lectureship at the London School of Economics in 1971, becoming professor in 1998.

Shing-Tung Yau (1949–). Chinese mathematician who has done highly influential work on differential geometry and partial differential equations. He is an analyst's geometer (or geometer's analyst) with enormous technical power and insight. He has cracked problems on which progress has been stopped for years. Yau proved the Calabi conjecture in 1976. Another conjecture proved by Yau was the positive mass conjecture, which comes from Riemannian geometry.

James A. Yorke (1941–). U.S. mathematician whose 1975 paper with T. Y. Li introduced the term "chaos." This originally referred to iterations which eventually are periodic with all periods. Later it came to be a more general term, encompassing evolutions that have "strange attractors," and evolutions in which even very small perturbations of their initial conditions can produce very large effects eventually.

Gail S. Young (1915–1999). U.S. mathematician who was a student of Robert Lee Moore at the University of Texas. Young held appointments at Purdue, Michigan, Tulane, Rochester, Wyoming, and Columbia and was chair of mathematics at two of these. He was also a president of the Mathematical Association of America. Moore was famous for his version of student-centered mathematics "instruction," and Young was one of his many students who became very prominent as researchers and organization officials.

Grace Chisholm Young (1868–1944). English mathematician who, with William Young, wrote 220 mathematical articles and several books. It is almost impossible to tell exactly how much of the work in these papers was due to Grace Young. As William Young wrote himself, "I feel partly as if I were teaching you, and setting you problems which I could not quite do myself but could enable you to."

William Henry Young (1863–1942). English mathematician who discovered Lebesgue integration, independently but 2 years after Lebesgue. He studied Fourier series and orthogonal series in general.

Oscar Zariski (1899–1986). Ukrainian-American professor of mathematics at Harvard who worked on the foundations of algebraic geometry using algebraic methods. He contributed to the theory of normal varieties, local uniformization, and the reduction of singularities of algebraic varieties.

Doron Zeilberger (1950–). Israeli-American professor of mathematics at Rutgers Unniversity who has made numerous important contributions to combinatorics, hypergeometric identities, and q-series.

Together with Herbert Wilf, Zeilberger was awarded the AMS Steele Prize for their development of WZ theory, which has revolutionized the field of hypergeometric summation. Zeilberger is known as a champion of using computers and algorithms to do mathematics quickly and efficiently. He credits his computer "Shalosh B. Ekhad" as a coauthor. ("Shalosh" and "Ekhad" mean "Three" and "One" in Hebrew respectively, referring to the AT&T 3B1 model).

Andrei Zelevinsky. Russian-American professor of mathematics at Northeastern University, researcher in algebra and combinatorics, and instructor at the Jewish People's University.

Notes

Introduction

1. Immordino-Yang M. H., & Damasio A. (2007). We feel, therefore we learn: The relevance of affective and social neuroscience to education. *Mind, Brain, and Education* 1(1), 2, 4.

2. Ibid., p. 10.

3. The concept of myth is included in Claudia Henrion's book *Women in Mathematics* as well as in this one. We use this notion independently of each other. It was only recently that we became aware of our shared use of this concept and some similarities in Henrion's and our formulations.

4. Bronowski, J. (1965). *Science and human values*. New York: Harper and Row, pp. 62–63.

Chapter 1. Mathematical Beginnnings

1. A short biographical list is included at the end of the book describing key achievements and facts about each of the mathematicians whose names appear in the text.

2. Ulam, Stan. (1976). *Adventures of a mathematician*. New York: Scribner, p. 10.

3. Dyson, Freeman J. (2004). Member of the club. In John Brockman (Ed.). *Curious minds: How a child becomes a scientist*. New York: Pantheon Books, p. 61.

4. Sacks, Oliver (2001). *Uncle Tungsten: Memories of a chemical boyhood*. New York: Alfred Knopf, p. 26.

5. Murray, Margaret (2000). *Women becoming mathematicians*. Cambridge, Mass.: Massachusetts Institute of Technology. Includes each of the mathematicians whose names appear in the text. Ulam (1976), pp. 10, 83.

6. Wigner, Eugene (1992). *The recollections of Eugene P. Wigner as told to Andrew Szanton*. New York: Plenum Press, p. 23.

7. Ibid., p. 45.

8. Ibid., pp. 47–48.

9. Brockman, John (Ed.) (2004). *Curious minds: How a child becomes a scientist*. New York: Pantheon Books, pp. 193–194.

10. Albers, Don (2007). John Todd—Numerical mathematics pioneer, *College Mathematics Journal* 38(1), 5.

11. Ibid.

12. Ibid.

13. Reid, Constance (1996). *Julia, a life in mathematics*. Washington, D.C.: Mathematical Association of America, p. 3.

14. Perl, Teri (1978). Biographies of women mathematicians and related activities. Menlo Park, Calif.: Addison-Wesley, p. 64.

15. Osen (2004).

16. James, I. (2002). *Remarkable mathematicians*. Cambridge: Cambridge University Press, pp. 231–232.

17. Ibid.

18. Ibid., p. 122.

19. Massera, J. L. (1998). *Recuerdos de mi vida academica y politica* (Memories of my academic and political life). Lecture delivered at the National Anthropology Museum of Mexico City, March 6, 1998, and published in *Jose Luis Massera: The scientist and the man*. Montevideo, Uruguay: Faculty of Engineering. Text translated by Frank Wimberly. These quotes are from the unpaginated text of Massera's lecture.

20. Mordell, L. (1971). Reminiscences of an octogenarian mathematician, *American Mathematical Monthly 78*, 952–961.

21. Tikhomirov, V. M. (2000). Moscow mathematics 1950–1975. In Jean Paul Pisier (Ed.). *Development of mathematics 1950–1975*. Boston: Birkhäuser, pp. 1109–1110.

22. Ibid.

23. Singh, Simon (1998). *Fermat's Last Theorem*. London: Fourth Estate, pp. 5–6.

24. Ibid., p. 6.

25. Feldman, David H. (1986). *Nature's gambit*. New York: Basic Books, p. 16.

26. Wolpert, Stuart (2006). Terence Tao, "Mozart of Math," is UCLA's first mathematician awarded the Fields Medal, often called the "Nobel Prize in Mathematics." *UCLA News*, August 22, 2006. Retrieved April 10, 2008, from http://newsroom.ucla.edu/portal/ucla/Terence-Tao-of-Math-7252.

27. Chang, Kenneth (2007). Journeys to the distant fields of prime. *New York Times*, March 13, 2007. Retrieved April 10, 2008 from http://www.nytimes.com/2007/03/13/science/13prof.html?_r=1&sq=The%20Mozart%20of%20MAth&st=nyt&oref=slogin&scp=1&pagewanted=print.

28. Winner, Ellen (1996). *Gifted children: Myths and realities*. New York: Basic Books, pp. 36–37.

29. Radford, John (1990). *Child prodigies and exceptional achievers*. New York: Harvester Wheatsheaf, p. 82.

30. Gustin, William C. (1985). The development of exceptional research mathematicians. In Benjamin S. Bloom (Ed.). *Developing talent in young people*. New York: Ballantine Books, p. 274.

31. Ibid., p. 279.

32. Winner (1996), p. 187.

33. Rathunde, Kevin, & Csikszentmihalyi, Mihaly. (1993). Undivided interest and the growth of talent: A longitudinal study of adolescents. *Journal of Youth and Adolescence* 22(4), 385–405.

34. Murray (2000), p. 49.

35. Paulson, Amanda. (2004). Children of immigrants shine in math, science, *Santa Fe New Mexican* 813 1/04, p. A5. These findings are further supported in a 2008 article in *Notices of the American Mathematical Society*.

36. Olson, Steve (2004). *Count down*. Boston: Houghton Mifflin, p. 63.

37. Heims, Steve J. (1982). *John von Neumann and Norbert Wiener: From Mathematics to the Technologies of Life and Death*. Cambridge, Mass.: MIT Press, p. 56.

38. Wiener, Norbert (1953). *Ex-prodigy: My childhood and youth*. New York: Simon and Schuster, pp. 67–68.

39. Levinson, N. (1966). Wiener's life, *Bulletin of the American Mathematical Society* 72(1, II), 3.

40. MacTutor, web site. George Dantzig, Birkhoff, p. 1.

41. Gustin (1985), pp. 279–282.

42. Feldman (1986), p. 12.

43. Howe, Michael J. A. (1990). *The origins of exceptional abilities*. Cambridge, Mass.: Basil Blackwell, p. 181.

44. Feldman (1986), p. 31.

45. Gustin (1985), p. 277.

46. Feldman (1986), p. 31.

47. Ibid., p. 33.

48. Ibid., p. 33.

49. Winner (1996), p. 19.

50. Ibid., p. 210.

51. Radford (1990), p. 96.

52. Gustin (1985), p. 278.

53. Ibid., p. 287.

54. Wigner (1992), p. 50.

55. Murray (2000), p. 67.

56. Ibid., p. 79.

57. Morrow, Charlene, & Perl, Teri (Eds.). (1998). *Notable women in mathematics: A biographical dictionary*. Westport, Conn.: Greenwood Press, pp. 190–191.

58. Ibid., p. 192.

59. Ibid., p. 193.

60. Murray (2000), p. 114.

61. Gallian, Joseph A. (2004). A conversation with Melanie Wood. *Math Horizons* 12 (September 2004), 123.

62. Olson (2004), p. 30.

63. Gallian (2004), p. 13.

64. Hersh, Reuben, & John-Steiner, Vera. (1993). A visit to Hungarian mathematics. *Mathematical Intelligencer* 15(2), 13–26.

65. Bloom (1985), pp. 308–309.

66. Notes on the theory and application of Fourier transforms, with R.E.A.C. Paley, *Transactions of the American Mathematical Society 35*.

67. Albers, Donald J., & Alexanderson, G. L. (1985). *Mathematical people: Profiles and interviews*. Boston: Birkhäuser, p. 287.

68. Ulam (1976), pp. 25–26.

69. Albers & Alexanderson (1985), p. 125.

70. O'Connor, J.J., & Robertson, E. F. (2003). George Dantzig, p. 2.

71. Hersh & John-Steiner (1993), p. 17.

72. Ibid., p. 18.

73. Yau, S. T. (Ed.) (1998). *S. S. Chern: A great geometer of the twentieth century*. Singapore, International Press.

74. Ibid.

75. Ibid.

76. Murray, Margaret (2000). *Women becoming mathematicians*, Cambridge, Mass.: Massachusetts Institute of Technology, p. 149.

77. Reid, C. (1976). *Courant in Göttingen and New York*. New York: Springer-Verlag, p. 255.

78. Hersh & John-Steiner (1993).

79. MacTutor, Birkhoff.

80. "Distant teachers" refers to significant teachers from the past whose work was influential to the novice without their ever meeting. This term comes from the work of Vera John-Steiner.

81. John-Steiner, Vera (1997). *Notebooks of the mind: Explorations of thinking*. New York: Oxford University Press, p. 54.

82. Gustin (1985), p. 326.

83. Ibid.

84. Ibid., p. 329.

Chapter 2. Mathematical Culture

1. Albers & Alexanderson (1985), p. 127.

2. Schwartz, L. (2001). *A mathematician grappling with his century*. Basel: Birkhäuser, p. 38.

3. Peter, R. (1990). Mathematics is beautiful. *Mathematical Intelligencer* 12(1), p. 58.

4. Boas, R. (1990). Interview. In D. J. Albers, G. L. Alexanderson, & C. Reid (Eds.). *More mathematical people: Contemporary conversations*. Boston: Birkhauser, p. 25.

5. Gardner, H. (1993). *Creating minds*. New York: Basic Books.

6. Everett, D. (2005). Cultural constraints on grammar and cognition in Piraha. *Current Anthropology* (Aug./Sept.), pp. 622–623.

7. Mandelbrot, B. (1985). Interview. In D. J. Albers & G. L. Alexanderson (Eds.), *Mathematical people: Profiles and interviews*. Boston: Birkhäuser, p. 209.

8. Ibid. p. 210.

9. Davis, P. J., & Hersh, R. (1981). *The mathematical experience*. Boston: Birkhäuser, p. 391.

10. Hadamard, J. (1945). *The psychology of invention in the mathematical field*. New York: Dover, p. 140.

11. Ibid., p. 142.

12. Weil, A. (1992). *The apprenticeship of a mathematician*. Basel: Birkhäuser, p. 91.

13. Csikszentmihalyi, M. (1990). *Flow: The psychology of optimal experience*. New York: Harper Perennial, p. 53.

14. Parikh, C. (1991). *The unreal life of Oscar Zariski*. Boston: Academic Press, p. 50.

15. Ibid., p. 51.

16. Atiyah, M. (1984). Interview. *Mathematical Intelligencer* 6(1), 17.

17. Macrae, Norman. (1992). *John von Neumann*. New York: Random House, pp. 44-51.

18. Gregory, Graves, & Bangert, N. (2005). *Math with heart: Why do mathematicians love math?* Unpublished manuscript.

19. Chang (2007), p. D1.

20. Davis, P. J., & Hersh, R. (1981). *The mathematical experience*. Boston: Birkhäuser, p. 310.

21. Ibid., p. 311.

22. Vygotsky, L. S. (1962). *Thought and language*. Cambridge, Mass.: MIT Press.

23. Rota, G. C. (1997). *Indiscrete thoughts*. Boston: Birkhäuser, pp. 45–46.

24. Bell, E. T. (1965). *Men of mathematics*. New York: Simon and Schuster, p. 378.

25. Ruelle, D. (2007). *The mathematician's brain*. Princeton, N.J.: Princeton University Press, p. 129.

26. Ibid., p. 96.

27. Ibid.

28. Hardy, G. H. (1991). *A mathematician's apology*. Cambridge: Cambridge University Press, p. 85.

29. Halmos, P. R. (1985). *I want to be a mathematician*. New York: MAA Spectrum, Springer-Verlag, p. 51.

30. Ibid., p. 73.

31. Mozzochi, C. J. (2000). *The Fermat diary*. Providence, R.I.: American Mathematical Society, pp. 64–65.

32. Singh (1998), p. 304.

33. da C. Andrade, E. N. (1954). *Sir Isaac Newton: His life and work*. Garden City, N.Y.: Anchor Books, p. 35.

34. Ibid.

35. Krantz, S. G. (2002). *Mathematical apocrypha: Stories and anecdotes of mathematicians and the mathematical.* Washington, D.C.: Mathematical Association of America, pp. 24–25.

36. Krantz, S. G. (2005). *Mathematical apocrypha redux: More Stories and anecdotes of mathematicians and the mathematical.* Washington, D.C.: Mathematical Association of America, p. 74.

37. Krantz (2002), p. 38.

38. Nasar, S., & Gruber, D. (2006). *Manifold destiny. The New Yorker,* August 28, 2006, p. 52.

39. Ibid., p. 45.

40. Collins, M. A., & Amabile, T. M. (1998). Creativity and motivation. In R. J. Sternberg (Ed.). *Handbook of creativity.* Cambridge: Cambridge University Press, p. 298.

41. Sternberg, J., & Lubart, T. I. (1991). The investment theory of creativity and its development. *Human Development 34,* 1–31.

42. Diacu F., & Holmes, P. (1996). *Celestial encounters.* Princeton, N.J.: Princeton University Press, p. 42.

43. Barrow-Green, J. (1997). *Poincaré and the three body problem.* Providence, R.I.: American Mathematical Society, p. 162.

44. Mira, C. (2000). I. Gumowski and a Toulouse research group in the "prehistoric" times of chaotic dynamics. In R. Abraham and Y. Ueda (Eds.). *The chaos avant-garde: Memories of the early days of chaos theory.* Singapore: World Scientific, p. 188.

45. Ueda, Y. (2000a). Strange attractors and the origins of chaos. In R. Abraham & Y. Ueda (Eds.). *The chaos avant-garde: Memories of the early days of chaos theory.* Singapore: World Scientific, p. 34.

46. Ueda, Y. (2000b). My encounter with chaos. In R. Abraham & Y. Ueda (Eds.). *The chaos avant-garde: Memories of the early days of chaos theory.* Singapore: World Scientific, pp. 48–49.

47. Ueda, Y. (2000c). Reflections on the origin of the broken-egg chaotic attractor. In R. Abraham & Y. Ueda (Eds.). *The chaos avant-garde: Memories of the early days of chaos theory.* Singapore: World Scientific, p. 65.

48. Abraham (2000), p. 89.

49. See Allyn Jackson's article in *Notices of the American Mathematical Society* which appeared in 1994.

50. Jackson, A. (1994). Fighting for tenure: The Jenny Harrison case opens a Pandora's box of issues about tenure, discrimination, and the law. *Notices of the American Mathematical Society 41*(3), p. 187.

51. John-Steiner, V. (2006). Harrison interview, December 4, 2006, Berkeley, Calif.

52. Harrison, J. (2007), web site.

53. Ibid.

54. Ibid.

55. Jackson (1994), p. 190.

56. Moore (2007). *Mathematics at Berkeley: A history.* Wellesley, Mass.: A. K. Peters, p. 288.

CHAPTER 3. MATHEMATICS AS SOLACE

1. Rota, G. C. (1990). "The lost café," in *Indiscrete Thoughts*. Boston: Birkhäuser, p.

2. Bollobas, Béla (Ed.). (1986). *Littlewood's miscellany*. Cambridge, Cambridge University Press.

3. "Rogers-Ramanujan identities" are identities between sums and products of rational functions, independently discovered by L. J. Rogers and Srinivasa Ramanujan.

4. Dyson, F. J. (1988). A walk through Ramanujan's garden. In G. E. Andrews et al. *Ramanujan revisited*. Boston: Academic Press, p. 15.

5. Weil's autobiography contains a myth about how the Finnish function theorist Rolf Nevanlinna saved his life. Nevanlinna claimed that when Weil was under arrest in Finland, Nevanlinna was at a state dinner also attended by the chief of police, and that when coffee was served the chief of police came to Nevanlinna saying: "Tomorrow we are executing a spy who claims to know you. Ordinarily I wouldn't have troubled you with such trivia, but since we're both here anyway I'm glad to have the opportunity to consult you." "What is his name?" "Andre Weil." Upon hearing this, Nevanlinna told Weil he was shocked. "I know him," he told the police chief. "Is it really necessary to execute him?" "Well, what do you want us to do with him?" "Couldn't you just escort him to the border and deport him?" "Well, there's an idea; I hadn't thought of it." "Thus," Weil wrote, "was my fate decided." We now retract this tale, which we innocently repeated in a version of this chapter published in the *Mathematical Intelligencer*. After that publication we belatedly learned that Nevanlinna's self-glorifying fairy tale was refuted in 1992 by the Finnish mathematician Osmo Pekonen. Pekonen read Weil's dossier in the Finnish police archives and found that Weil was never condemned to death, and Nevanlinna was never involved in his case. (In an interview in the United States in 1934, Hermann Weyl called Nevanlinna a "Finnish Nazi." In World War II Nevanlinna served as chairman of the Committee for the Finnish Volunteer Battalion of the Waffen-SS.) [Osmo Pekonen, *L'affaire Weil à Helsinki en 1939*. *Gazette des mathématiciens 52* (April 1992), pp. 13–20.]

6. Weil (1992).

7. Massera (1998).

8. Each day, for some hours, each prisoner could have tools in his cell, that were delivered by an inmate, with the objective of making handicrafts. The sale of these, during the vexing and frustrating work toward release, helped with family sustenance.

9. The *Isla* (island) was a place of punishment where the prisoner stayed in a state of complete isolation; the only thing within the walls was his body. At night they delivered a blanket and quilt. And three times a day a soldier, with whom he was prohibited communication, delivered a meager food ration. It was the coldest, least hospitable place in the prison.

10. The entry of books written in languages besides Spanish was always prohibited. And in the epoch of which we write, every mathematics book

was prohibited from entering: they might have codes that the censor couldn't interpret.

11. Abramowitz, I. (1946). *The great prisoners*. New York: E. P. Dutton, p. 142.

12. Turán, P. (1997). Note of welcome. *Journal of Graph Theory 1*(1), 1.

CHAPTER 4. MATHEMATICS AS AN ADDICTION:
FOLLOWING LOGIC TO THE END

1. Schneps, L. (2008). *Grothendieck-Serre correspondence*, book review. *Mathematical Intelligencer 30*(1), 66–68.

2. Roy Lisker translated the first 100 pages of *Récoltes et Semailles* (Reaping and Sowing) and published it in his magazine *Ferment*.

3. Ibid.

4. Ibid.

5. Ibid.

6. Schwartz (2001).

7. Ibid.

8. Ibid.

9. Grothendieck, A. (1986). *Recoltes et Semailles*, unpublished manuscript, promenade #9.

10. Grothendieck, A., Colmez, P. (Ed.), & Serre, J. P. (2001). *Grothendieck-Serre correspondence*. Paris: Societe Mathematique Francaise. Bilingual edition, Providence, R.I.: American Mathematical Society, 2004.

11. Grothendieck (1986), promenade #10.

12. Ibid., promenade #11.

13. Ibid., promenade #13.

14. Ibid., promenade, #7.

15. Ruelle, D. (2007). *The mathematician's brain*. Princeton, N.J.: Princeton University Press, p. 40.

16. Ibid.

17. Grothendieck, A. (1989). Letter refusing the Crafoord Prize, *Le Monde*, May 4, 1998. *Mathematical Intelligencer 11*(1), 34–35.

18. Grothendieck (1986), promenade #17.

19. The unsettling story was reported in the *Ukrainian Weekly* on May 12, 1996, and August 31, 1997, and is available on the World Wide Web.

20. Baruk, H. (1978). *Patients are people like us*. New York: William Morrow, pp. 184–187.

21. Ibid.

22. Ibid.

CHAPTER 5. FRIENDSHIPS AND PARTNERSHIPS

1. Bollobás (1986), p. 8.

2. Koblitz, A. (1983). *A convergence of lives*. Boston: Birkhäuser, pp. 99–101.

3. Ibid., p. 101.

4. Ibid., p. 113.

5. James (2002), pp. 233–234.

6. Ibid.

7. Reid, C. (2004). *Hilbert*. New York: Springer, pp. 12–13.

8. Ibid., p. 14.

9. Ibid., p. 91.

10. Ibid., p. 95.

11. Ibid.

12. Ibid.

13. Hardy, G. H. (1967). *A mathematician's apology*. New York: Cambridge University Press, pp. 147, 148.

14. Snow, C. P. (1967). Foreword to Hardy (1967). *A mathematician's apology*. New York: Cambridge University Press, p. 29.

15. Bollobás (1986), p. 2.

16. Ibid., p. 11.

17. Tattersall, J., & McMurran, S. (2001). *An interview with Dame Mary L. Cartwright*, D. B. E., F. R. S. *College Mathematics Journal 32*(4), 249.

18. Bollobás (1986), p. 13.

19. URL in references.

20. Tattersall & McMurran (2001), p. 249.

21. Hardy, G. H. (1978). *Ramanujan*. New York: Cambridge University Press, pp. 2–3.

22. Kanigel, R. (1991). *The man who knew infinity*. New York: Simon and Schuster, p. 294.

23. Hardy (1967), p. 148.

24. Snow, C. P. (1967), p. 51.

25. Ibid.

26. Ibid., p. 57.

27. Aleksandrov, P. S. (2000). "A few words on A. N. Kolmogorov," *Russian Mathematical Surveys 39*(4), 5–7, in *Kolmogorov in perspective*. Providence, R.I.: American Mathematical Society, p. 142.

28. Kolmogorov, A. N. (2000). "Memories of P. S. Aleksandrov" (Russian Mathematical Surveys *41*, 255–246), in *Kolmogorov in perspective*. Providence, R.I.: American Mathematical Society, p. 145.

29. Ibid., p. 150.

30. Ibid., p. 156.

31. Ibid., p. 152.

32. Ibid., p. 159.

33. Gessen, Masha (2005). *Perfect rigor*. New York: Houghton Mifflin Harcourt, p. 43.

34. Ulam (1976), pp. 33–34.

35. Heims (1982), p. 42.

36. Abelson, P. (1965). Relation of group activity to creativity in science. *Daedalus* (summer 1965), p. 607.

37. Rota, G. C. (1987). The lost café, *Los Alamos Science*, Special Issue, 26, p. 26.

38. Schechter, B. (1998). *My brain is open: The mathematical journeys of Paul Erdős*, New York: Simon and Schuster, p. 196.

39. Goldstein, R. (2006). *Incompleteness: The proof and paradox of Kurt Gödel (Great discoveries)*. New York: W. W. Norton, p. 29.

40. Ibid., p. 33.

41. Ibid.

42. Grattan-Guinness, Ivor. (1972). A mathematical union *Annals of Science 29*(2), 105–186.

43. Ibid., p. 115.

44. Ibid., p. 118.

45. Ibid., p. 121.

46. Ibid., p. 123.

47. Ibid., p. 131.

48. Ibid., pp. 131–132.

49. Ibid., p. 140.

50. Ibid., p. 141–142.

51. Ibid., p. 148.

52. Ibid., p. 149.

53. Ibid., p. 151.

54. Ibid., p. 156.

55. Ibid.

56. Ibid., p. 163.

57. Ibid., p. 166.

58. Ibid., p. 177.

59. Ibid., p. 181.

60. Reid (1996), p. 35.

61. Ibid., p. 38.

62. Ibid., p. 37.

63. Ibid., p. 43.

64. Ibid., p. 45.

65. Ibid., p. 79.

66. Ibid.

67. Ibid., 39, 79.

68. Tausky-Todd, O. (1985). Autobiographical Essay. In D. J. Albers and G. L. Alexanderson (Eds.). *Mathematical people: Profiles and interviews*. Boston: Birkhauser, p. 321.

69. Ibid., p. 325.

70. Birman, J. et al. (1991).

71. Helson, R. (2005). Personal communication.

72. Case, B. A. & Leggett, A. M. (Eds.) (2005). *Complexities: Women in mathematics*. Princeton, N.J.: Princeton University Press, p. 189.

73. Birman, J. et al. (1991). p. 702.

74. Case, B. A., et al., p. 190.

75. Gardner (1993), p. 385.

Chapter 6. Mathematical Communities

1. Ulam (1976), p. 38.

2. Rowe, D. E. (1989). Klein, Hilbert, and the Göttingen mathematical tradition. In K. M. Oleska (Ed.). *Science in Germany: The intersection of institutional and intellectual issues. Osiris 5*, 189–213.

3. Gowers, T., & Nielson, M. (2009). Massively collaborative mathematics. *Nature 461*, 879–881.

4. Albers & Alexanderson (Eds.) (1985).

5. Beaulieu (1993). A Parisian café and ten proto-Bourbaki meetings (1934–1935). *Mathematical Intelligencer 15*(1), 27–35.

6. Dieudonné, J. A. (1970). The work of Nicholas Bourbaki. *American Mathematical Monthly 77*, 134–145.

7. Ibid.

8. Grothendieck (1986).

9. Cartan, H. M. (1980). Nicolas Bourbaki and contemporary mathematics. *Mathematical Intelligencer 2*(4), 175–187.

10. Klein, F. (1979). Development of mathematics in the 19th century. Translated by M. Ackerman. In R. Hermann (Ed.). *Lie groups, history, frontiers and applications,* vol. IX. Brookline, Mass.: Math Science Press.

11. Courant, R. (1980). Reminiscences from Hilbert's Göttingen. *Mathematical Intelligencer 3*(3), 159.

12. Honda, K. (1975). Teiji Takagi: A biography. Commentary. *Mathematica Universitatis Sancti Pauli* XXIV-2, 141–167.

13. Alexandrov (2000).

14. Lewy, Hans (1992). Quoted by F. John.

15. John, F. (1992). Memories of student days in Göttingen. *Miscellanea Mathematica,* New York: Springer, pp. 213–220.

16. Courant, R. (1980). Reminiscences from Hilbert's Göttingen. *Mathematical Intelligencer 3*(3), 163–164.

17. Courant's eulogy for Friedrichs.

18. Moser, J. (1995). Obituary for Fritz John, 1910–1994. *Notices of the American Mathematical Society 42*(2), 256–257.

19. Lui, S. H. (1997). An interview with Vladimir Arnol'd. *Notices of the American Mathematical Society 42*(2), 432–438.

20. Sossinsky, A. B. (1993). In the other direction. In S. Zdravkovska & P. L. Duren (Eds.). *Golden years of Moscow mathematics, History of mathematics,* vol. 6. Providence, R.I.: American Mathematical Society, pp. 223–243.

21. Ibid.

22. Fuchs, D. B. (1993). On Soviet mathematics of the 1950s and 1960s. In S. Zdravkovska & P. L. Duren (Eds.). *Golden years of Moscow mathematics, History of mathematics,* vol. 6. Providence, R.I.: American Mathematical Society, pp. 220–222.

23. Ibid.

24. Sossinsky (1993), pp. 223–243.

25. Ibid.

26. Ibid.

27. Ibid.

28. Zelevinsky, A. (2005). Remembering Bella Abramovna. In M. Shifman (Ed.). *You failed your math test, Comrade Einstein.* Hackensack, N.J.: World Scientific.

29. Henrion, C. (1997). *Women in mathematics.* Bloomington, Ind.: Indiana University Press, p. 152.

30. *Notices* of the *American Mathematical Society* (1997), p. 107.

31. Lewis, D. J. (1991). Mathematics and women: The undergraduate school and pipeline. *Notices of the American Mathematical Society* 38(7), 721–723.

32. Roitman, J. (2005). In B. A. Case and A. M. Leggett (Eds.). *Complexities: Women in Mathematics.* Princeton, N.J.: Princeton University Press, p. 251.

CHAPTER 7. GENDER AND AGE IN MATHEMATICS

1. Marjorie Senechal (2007). Hardy as mentor. *Mathematical Intelligencer* 29(1), 16–23.

2. Germain-Gauss correspondence <http://www-groups.dcs.st-and.ac.uk/%7Ehistory/Mathematicians/Gauss.html>

3. LaGrange <http://www-groups.dcs.st-and.ac.uk/%7Ehistory/Mathematicians/Lagrange.html>

4. James (2002), pp. 57–58.

5. Kovalevskaya, S., Kochina, P. Y., & Stillman, B. (1978). *A Russian childhood.* New York: Springer, p. 35.

6. *Crelle's Journal* <http://www-groups.dcs.st-and.ac.uk/%7Ehistory/Mathematicians/Crelle.html>

7. Weierstrass <http://www-groups.dcs.st-and.ac.uk/%7Ehistory/Mathematicians/Weierstrass.html>

8. Ibid.

9. Kovalevskaya et al. (1978), p. 35.

10. James (2002), p. 235.

11. Smolin, L. (2006). *The trouble with physics.* London: Houghton Mifflin Penguin.

12. Reid, C. (1986). *Hilbert-Courant.* New York: Springer-Verlag, p. 143.

13. Weyl, H. (1935). Emmy Noether. *Scripta Mathematica* 3(3), 201–220.

14. James, I. (2009). *Driven to innovate: A century of Jewish mathematicians and physicists.* Oxford: Peter Lang.

15. *Notices of the American Mathematical Society* of America, 2005.

16. Henrion (1997), p. xvii.

17. Ibid., p. 66.

18. Ibid., p. 18.

19. Ibid., p. 44.

20. Ibid., p. 73.

21. Ibid., p. 134.

22. Albers, D., & Alexanderson, G. (1991). A conversation with Ivan Niven. *College Mathematics Journal* 22(5), 371–402.

23. Ibid., p. 393.

24. Ibid.

25. Hyde, J. (2005). The gender similarities hypothesis. *American Psychologist 60*, 581–592.

26. Henrion (1997), p. 208.

27. Ibid., p. 134.

28. Bollobás (1986), pp. 15–16.

29. Personal communication.

30. Mordell's name is usually mentioned today in connection with his conjecture of 1922, finally proved by Gerd Faltings of Germany in 1983: A smooth rational plane curve of genus greater than 1 has finitely many points with rational coefficients.

31. Mordell, L. J. (1970). Hardy's *A mathematician's apology. American Mathematical Monthy* 77, 836.

32. Ibid.

33. Ibid.

34. Ibid.

35. Einstein (1942), p. 150.

36. Weil, A. (1950). The future of mathematics. *American Mathematical Monthly 57*, 296.

37. Wiegand, S. (1977). Grace Chisholm Young. *Association for Women in Mathematics Newsletter 7*, 6.

38. Adler, A. (1972). Mathematics and creativity. *New Yorker*, February 19, 1972. Reprinted in Timothy Ferris (Ed.) (1993). *The world treasury of physics, astronomy, and mathematics*. Back Bay Books, p. 435.

39. van Stigt, W. P. (1990). *Brouwer's intuitionism*. Amsterdam: North-Holland.

40. Dauben, J. (1995). *Abraham Robinson: The creation of nonstandard analysis: A personal and mathematical odyssey*. Princeton, N.J.: Princeton University Press, p. 491.

41. Schneider, Ivo, e-mail communication.

42. Bell (1937), p. 405.

43. An earlier version of this survey appeared as Hersh, R. (2001). Mathematical menopause, or, a young man's game? *Mathematical Intelligencer* 23(3), 52–60.

44. Taylor, S. S. (1999). *Research dialogues of the TIM-CREF*, no. 62.

45. Henrion, p. 113.

46. Henrion, C. (1997).

47. Henrion (1997), p. 112.

48. Ibid.

49. Bollobás (1986), p. 14.

50. Personal communication.

51. Bollobás (1986), p. 14.

CHAPTER 8. THE TEACHING OF MATHEMATICS: FIERCE OR FRIENDLY

1. Albert Einstein. Quoted in Holton, G. (1973). *Thematic origins of scientific thought: Kepler to Einstein.* Cambridge, Mass.: Harvard University Press, p. 377.

2. Parker, J. (2004). *R. L. Moore: Mathematician and teacher.* Washington, D.C.: Mathematical Association of America, p. 3.

3. Megginson, R. E. (2003). Yueh-Gin Gung and Dr. Charles Y. Hu award to Clarence F. Stephens for distinguished service to mathematics. *American Mathematical Monthly* 110(3), 177–180.

4. Mathematicians of the African Diaspora web site.

5. Megginson (2003), p. 177.

6. Donaldson (1989), p. 450.

7. *Dictionary of American Biography*, p. 163.

8. Parker (2004), pp. vii, viii.

9. Moore specialized in what became known as "Moore spaces." These are topological spaces satisfying a technical additional condition that was specified in part 4 of Moore's Axiom 1. [Wilder, R. L. (1982). The mathematical work of R. L. Moore: Its background, nature and influence. *Archive for the History of Exact Sciences* 26, 73–97]. A *"complete* Moore space" satisfied all four parts of Axiom 1. According to Wilder, over 300 papers had then been published on the question, When is a Moore space metrizable? In 1951 Moore's student R H Bing proved that a Moore space is metrizable if it is "collectionwise normal."

10. Parker (2004), p. 244.

11. Albers, D. J. (1990). *More mathematical people.* Boston: Harcourt Brace Jovanovich, p. 293.

12. Parker (2004), p. 271.

13. Ibid.

14. Ibid., p. 226.

15. Mathematicians of the African Diaspora web Site.

16. Ibid.

17. Parker (2004), p. 288.

18. Personal communication.

19. Ibid.

20. Parker (2004), p. 288.

21. Personal communication (2007).

22. Ibid.

23. Claytor gave a necessary and sufficient condition for a Peano continuum to be homeomorphic to a subset of the surface of a sphere, improving on earlier results by the famous Polish topologist Casimir Kuratowski.

24. Ibid.

25. Ibid.

26. Albers & Alexanderson (1985), p. 23.

27. Megginson (2003), p. 179.

28. Poland, J. (1987). A modern fairy tale? *American Mathematical Monthly 94*(3), 293.

29. Ibid.

30. Datta, D. K. (1993). Math education at its best: *The Potsdam model*. Framingham, Mass.: Center for Teaching/Learning of Mathematics, p. 5.

31. Ibid., pp. 65–66.

32. Ibid., p. 9.

33. Ibid.

34. Ibid., p. 23.

35. Personal communication (2006).

CHAPTER 9. LOVING AND HATING SCHOOL MATHEMATICS

1. Cornell, C. (1999). I hate math! I couldn't learn it, and I can't teach it! *Childhood Education 75*(4), p. 1.

2. Tobias, Sheila. (1993). *Overcoming math anxiety*. New York: W. W. Norton.

3. Lester, W. (2005). Hate mathematics? You are not alone. Associated Press, August 16, 2005.

4. Hardy, G. H. (1948). *A mathematician's apology*. New York: Cambridge University Press, pp. 86–87.

5. Slocum, J., & Sonneveld, D. (2006). The 15 puzzle. Beverly Hills, Calif., Slocum Puzzle Foundation, p. 9.

6. Personal communication (2006).

7. Cohen, R. (2006). What is the value of algebra? *Washington Post*, February 16, 2006.

8. McCarthy, C. (1991). Who needs algebra? *Washington Post*, April 20, 1991.

9. Ibid.

10. Russell, B. (1957). "The study of mathematics," in *Mysticism and logic*. New York: Doubleday, p. 60.

11. Ibid., p. 208.

12. Ibid., p. 68.

13. Zaslavsky, C. (1996), *The multicultural classroom*. Portsmouth, N.H.: Heinemann, p. 60.

14. Charbonneau M., & John-Steiner, V. (1988). Patterns of experience and the language of mathematics. In R. Cocking & J. P. Mestre (Eds.), *Linguistic and cultural influences on learning mathematics*. Hillsdale, N.J.: Erlbaum, p. 94.

15. Gerdes, P. (2001). On culture, geometrical thinking and mathematics education. In A. B. Powell & M. Frankenstein (Eds.). *Ethnomathematics, challenging Eurocentrism in mathematics education*. Albany, New York: State University of New York Press, pp. 231–232.

16. Carraher, T. N., Carraher, D., & Schliemann, A. D. (1985). Mathematics in the streets and in the schools. *British Journal of Developmental Psychology 3*, 21–29.

17. Schliemann (1995), p. 50.

18. Ibid.

19. PME 19, vol. 1, p. 20.

20. (1988), p. 44.

21. Ibid., p. 165.

22. Semisimple Lie groups are, . . .

23. Fosnot, C. T., & Dolk, M. (2001). *Young mathematicians at work*. Portsmouth, N.H.: Heinemann, p. 124.

24. Schmittau, Jean. (2003). Cultural historical theory in mathematics education. In Kozulin, A. Gindis, B. Ageyev V. S., Miller. S. N. (Eds.) *Vygotsky's Educational Theory in Cultural Context*. New York: Cambridge University Press.

25. Devlin, Keith. (2009). MAA On-line, January 2009. Should children learn math by starting with counting? Devlin's angle: http://www.maa .org/devlin/devlin_01_09.html

26. Umland, K. (2006). Personal communication.

27. Moses, R. P., & Cobb, C. E. Jr. (2001). *Radical equations: Math literacy and civil rights*. Boston: Beacon Press, pp. 10–11.

28. Ibid., p. 16.

29. Ibid., p. 146.

30. Ibid., p. 177.

31. Ibid., p. 179.

32. Ibid., p. 183.

33. Ibid., p. 18.

34. Treisman, V. (1992). Studying students studying calculus: A look at the lives of student mathematicians. *College Mathematics Journal 23*, p. 363.

35. Rota, G. C. (1997). *Indiscrete thoughts*. Boston: Birkhäuser, p. 39.

36. Levin, T. (2006). As math scores lag, a new push for basics. *New York Times*, November 14, 2006.

37. Ibid., p. 19.

38. Ibid.

39. Pearson, R. S. (1991). Why don't engineers use undergraduate mathematics in their professional work? *UME Trends 3*, 8.

40. Dudley, U. (1997). Is mathematics necessary? *College Mathematics Journal 28*(5), 361–365.

41. Krantz (2002), p. 61.

42. Halmos (1985), p. 261.

43. Noddings, N. (1993). Excellence as a guide to educational conversation. *Teachers College Record 94*(4), 8, 9.

44. Ibid., p. 13.

45. Noddings, N. (1994). Does everybody count? *Journal of Mathematical Behavior 13*(1), 10.

46. Ibid.

47. (1999), p. 32.

48. Umland, K. (2006). Personal communication.

Conclusions

1. Halmos (1985), p. 3.

2. Henrion (1997), p. 228.

3. Albers, D. J., Alexanderson, G. L., & Reid, C. (1990). *More mathematical people*. Boston: Harcourt Brace Jovanovich, pp. 3–26.

4. Ibid., p. 14.

5. Byers, W. (2007). *How mathematicians think: Using ambiguity, contradiction and paradox to create mathematics*. Princeton, N.J.: Princeton University Press, p. 78.

INDEX OF NAMES

GENERAL INDEX

DATE			
	WITHDRAWN		